Mary Somerville

In an era when science was perceived as a male domain, Mary Somerville (1780–1872) became both the leading woman scientist of her day and an integral part of the British scientific community. Her scientific writings contributed to one of the most important cultural projects of Victorian Britain: establishing science as a distinct, integral, and unifying element of culture. By the time of her death, Somerville had achieved near-mythic status in Britain. Her works reflect both the power of science to capture imagination and the influence of cultural factors in the development of science. They provide a window into a particularly lucid and illuminated mind and into one of the most formative periods in the evolution of modern scientific culture. This retelling of Somerville's story focuses on the factors that allowed her to become an eminent scientist and argues for rethinking the story of women's participation in science.

Kathryn A. Neeley is Associate Professor in the Division of Technology, Culture, and Communication at the University of Virginia; past President of the Humanities and Technology Association; and Chair of the Liberal Education Division of the American Society for Engineering Education. She has written for *Research in Philosophy and Technology*, *IEEE Transactions on Professional Communication*, and the proceedings of the American Society for Engineering Education. She is co-editor of the forthcoming *Liberal Education for 21st-Century Engineering*.

Cambridge Science Biographies

EDITORS
David Knight, *University of Durham*
Sally Gregory Kohlstedt, *University of Minnesota*

André-Marie Ampère, *Enlightenment and Electrodynamics*
JAMES R. HOFMANN
Charles Darwin, *The Man and His Influence* PETER J. BOWLER
Humphry Davy, *Science and Power* DAVID KNIGHT
Galileo, *Decisive Innovator* MICHAEL SHARRATT
Antoine Lavoisier, *Science, Administration and Revolution*
ARTHUR DONOVAN
Henry More, *and the Scientific Revolution* A. RUPERT HALL
Isaac Newton, *Adventurer in Thought* A. RUPERT HALL
Mary Somerville, *Science, Illumination, and the Female Mind*
KATHRYN A. NEELEY
Justus von Liebig, *The Chemical Gatekeeper* WILLIAM H. BROCK

MARY
SOMERVILLE
SCIENCE, ILLUMINATION, AND THE FEMALE MIND

KATHRYN A. NEELEY

CAMBRIDGE
UNIVERSITY PRESS

211 27

PUBLISHED BY THE PRESS SYNDICATE OF THE UNIVERSITY OF CAMBRIDGE
The Pitt Building, Trumpington Street, Cambridge, United Kingdom

CAMBRIDGE UNIVERSITY PRESS
The Edinburgh Building, Cambridge CB2 2RU, UK
40 West 20th Street, New York, NY 10011-4211, USA
10 Stamford Road, Oakleigh, VIC 3166, Australia
Ruiz de Alarcón 13, 28014 Madrid, Spain
Dock House, The Waterfront, Cape Town 8001, South Africa

http://www.cambridge.org

First published 2001

Printed in the United States of America

Typeface Palatino 10/12 pt.   *System* QuarkXPress [BTS]

A catalog record for this book is available from the British Library.

Library of Congress Cataloging in Publication Data
Neeley, Kathryn A. (Kathryn Angelyn), 1954–
Mary Somerville: science, illumination, and the female mind /
Kathryn A. Neeley.
p. cm. – (Cambridge science biographies series)
Includes bibliographical references and index.
ISBN 0-521-62299-9 – ISBN 0-521-62672-2 (pb)
1. Somerville, Mary, 1780–1872.   2. Women scientists – Great Britain –
Biography.   I. Somerville, Mary, 1780–1872.   II. Title.
III. Series.
Q143.S7 N44 2001
500.2'092 – dc21
[B]   00-049363

ISBN   0 521 62299 9   hardback
ISBN   0 521 62672 2   paperback

For the Choir Invisible

Contents

Author's Preface

In one of her best known poems, George Eliot wrote of "the choir invisible," the "immortal dead who live again / In minds made better by their presence." In the work of which this book is the culmination, I have been sustained and inspired by my own Choir Invisible, who have enlarged my view of what was possible and accompanied me on what has turned out to be one of the most challenging and rewarding journeys of my life. I dedicate this book to them.

The group has many members, all of whose voices have been important, but there are four who have been particularly strong presences in this work: my grandmothers, Nina Bookhardt Neeley and Jessie Rivers Von Harten, my great aunt Myrtle Von Harten Davison, and my dear friend Ann Doner Vaughan. Their capacity to inspire and support others has not been diminished by the passage from this life to the next.

The Choir also includes Mary Somerville, whom I have never properly met but who has been an excellent companion and most agreeable subject through the years I have spent with her. I have on occasion been accused of acting too much as her advocate. On one level, of course, this charge is quite justified. For as long as I have truly appreciated what she accomplished, I have felt an obligation to help bring it to light. An unusual and fortuitous set of circumstances put me in a position to comprehend her work, and I have worked hard to help others do so. I would add that I did not begin this project as Somerville's advocate. When I first encountered her, I saw Somerville and the esteem in which her contemporaries held her as a puzzle. It has only been with the passage of much time and the assistance of many generous people that

I have come close to understanding what she accomplished and why it was valued.

Another sustaining presence in this project has been Elizabeth Chambers Patterson, who catalogued the Somerville Collection, transformed a mountain of correspondence into a coherent account of Mary Somerville's career, and documented the eminence Somerville achieved in her own day. Without her work, mine simply would not have been possible, and I am very grateful to her.

I have also enjoyed the support of an extensive Choir Visible, who have helped me in more ways than I can possibly enumerate here. Mary Crum first introduced to me to the excitement of learning and to the numerous connections among literature, science, and life. Mel Cherno read the entire manuscript (some of it more than once) and offered both perceptive critique and cheerful encouragement.

Nina Byers opened my eyes to Mary Somerville's potential as a subject and introduced me to the resources of Somerville College. Pauline Adams, Hillary Ockendon, and Derek Goldrei, all of Somerville College, provided warm hospitality and many other forms of concrete aid. I am particularly grateful to Somerville College for permission to quote from the unpublished manuscripts of the Somerville Collection held at the Bodleian Library and for allowing me to use Mary Somerville's self-portrait as the frontispiece for this book. Simon Mitton, Alex Holzman, and Sally Gregory Kohlstedt of Cambridge University Press provided both inspiration and direction as the project evolved.

My dear friend Joe Vaughan died just as I was beginning the last round of revisions of this manuscript. To the end of his long life, he was a veritable choir in his own right. His generosity of spirit and capacity to help others recognize their better selves are qualities from which I have greatly benefited and which I hope to emulate. He read and critiqued this manuscript and offered the invaluable perspective of a pragmatic visionary.

Clare Hall and Cambridge University provided a wonderfully stimulating atmosphere that allowed me to refine my thinking about Somerville in ways I could not have previously imagined. I am grateful to Clare Hall's president, Gillian Beer, and to the fellows and other visitors at Clare Hall for their hospitality and willingness to share ideas.

I have enjoyed the support of numerous other friends and colleagues who mastered the fine art of asking "How is the book coming?" without conveying the least hint of anxiety. I would like in particular to mention Ralph and Libby Cohen, Mary Hamer, Nick Cumpsty, Susan Bernstein, Ingrid Townsend, Patricia Click, and George Webb, as well as Mike Bowles, who helped enormously with the extensive travel arrangements required for the completion of this project, and Mark Whittle, who read parts of the manuscript and offered an astronomer's point of view. I am also grateful to Amanda French and Johannah Mayfield, who served as my research assistants, and to Michael Dettelbach, who helped me find my way through the culture of nineteenth-century science.

But perhaps the most noteworthy group are those who did the most, who understood the enterprise the least, and whose involvement grew entirely out of their affection for me. They include my parents, Sam and Verna, and my sister, Betsi, into whose family it was my great good luck to be born, and my three sons, Michael, Arthur, and Elliott, each of whom, in his own distinctive way, has enriched my life beyond measure. But the greatest thanks are due to my husband, John Crafaik, whose strength and determination helped me create the space in my life where this work could be completed and who has made innumerable sacrifices for the development of my career. And, who knows, some day he may even read this book.

Charlottesville, Virginia
September 8, 1999

Key to Parenthetical References to the Works of Mary Somerville

Conn	*On the Connexion of the Physical Sciences* (1846 ed.)
Mech	*Mechanism of the Heavens* (1831)
M I	First manuscript draft of Somerville's autobiography, *Personal Recollections* (Somerville Collection)
M II	Second manuscript draft of *Personal Recollections* (Somerville Collection)
MMS	*On Molecular and Microscopic Science* (1869)
PD	*Preliminary Dissertation on Mechanism of the Heavens* (1832)
PG	*Physical Geography* (1850 ed.)
PR	*Personal Recollections from Early Life to Old Age of Mary Somerville* (1873) Note: Material that was included in M I or M II but omitted from the published version of *Personal Recollections* is distinguished in the text by italic type.

Note: The *Preliminary Dissertation on Mechanism of the Heavens* (1832) was originally published as an extended introduction to *Mechanism of the Heavens* (1831), then was published separately in 1832, and is often treated as a distinct work. The text is the same in both editions, but the pagination differs. The direct quotations from the "Preliminary Dissertation" that appear here are cited using the pagination of the 1831 text of *Mechanism of the Heavens*.

Prologue

Perceiving What Others Do Not Perceive

The "Peculiar Illumination" of the Female Mind

It is impossible to be a mathematician without being a poet in soul. . . . the poet has only to perceive that which others do not perceive, to look deeper than others look. And the mathematician must be able to do the same thing.

– Sónya Kovalévsky

Notwithstanding all the dreams of theorists, there is a sex in minds. One of the characteristics of the female intellect is a clearness of perception. . . . when women are philosophers, they are likely to be lucid ones; . . . when they extend the range of their speculative views, there will be a peculiar illumination thrown over the prospect.

– William Whewell,
Review of *On the Connexion of the Physical Sciences*
(1834) by Mary Somerville

Three large windows with extensive views dominate the room in which I wrote most of this book. As I immersed myself in the scientific writings of Mary Somerville and looked at the world outside through

1

the lenses her writing provided, those very familiar views took on an entirely new aspect, and the world of nature became much more vast, vivid, and dynamic than I had ever imagined it could be. My own powers of perception had been expanded as I watched Somerville exercise hers. I hope I have succeeded in conveying the power of Somerville's transforming vision. Its power derives in large part from Somerville's ability to use science to heighten perception and stimulate imagination.

The utilitarian goals commonly associated with science often overshadow its other functions. Natural processes tend to operate on a scale that is either too large or too small to be directly observed through our senses; yet, a scientifically informed imagination can transcend the limitations of the human senses, and language can be used to help imagination construct what cannot be directly observed. Similarly, our beliefs about the objectivity of scientific knowledge tend to obscure an important truth about our perceptions of nature. We tend to "see in Nature what we have been taught to look for" and to "feel what we have been prepared to feel." (Nicolson 1959: 5) Much of the cultural significance of science arises from the role it has played in teaching people to see and respond to the natural world.

This capacity to expand the effective perception of reality is shared by literature and science and is always grounded in imagination. When the Russian mathematician Sónya Kovalévsky wrote about the role of imagination in mathematics, she highlighted the ways in which science and mathematics fulfill functions usually ascribed to poetry and other forms of literature. She also called attention to the role of imagination and vision in all of these enterprises. Her analysis reminds us that the essence of any form of expertise is to see what others do not see – to go beyond surface appearances and the limitations of time and the senses to illuminate what would otherwise be hidden and to enlarge our powers of perception in the process. In describing her own experience, Kovalévsky emphasized the synergistic relationship of mathematics and literature in her intellectual life.

> As for myself, all my life I have been unable to decide for which I had the greater inclination, mathematics or literature. As soon as my brain grows wearied of purely abstract speculations it immediately begins to incline to observations on life, to narrative, and *vice versa*, everything in life begins to appear insignificant and uninteresting, and only the eternal, immutable laws of science attract me. It is very possible that I should have accomplished more in either of these lines, if I had devoted myself exclusively to it; nevertheless, I cannot give up either of them completely. (Kovalévsky quoted in Mozans 1913: 165–166)

Kovalévsky's description articulates what Somerville's scientific writing demonstrates, the rich world of a mathematician who is also "a poet in soul" and moves easily from abstract speculations to observations on life, from having her head among the stars to having her feet firm upon the earth. But Kovalévsky's analysis misses one very important point: both the mathematician and the poet must move beyond private insight to express what they perceive, to make it public and accessible to others. In Somerville's case, the capacities of the poet and the mathematician came together with those of the skillful writer who could not only help her readers see more but also help them see it much more clearly. This ability must have been what William Whewell was talking about when he described "the peculiar illumination" of the female mind. This quality of mind combines the perceptive power of science with that of poetry to go far beyond ordinary experience and to present a view that is at once precisely delineated, easy to comprehend, and pleasurable to contemplate.

Transforming Vision: Connecting Science, Gender, and Illumination

In the sonnet he wrote in response to *Mechanism of the Heavens* and published in his review of *On the Connexion of the Physical Sciences*, Whewell describes Somerville as one to whom "dark . . . seems bright, perplexed seems plain, / Seen in the depths of a pellucid mind." (68) One of the central characteristics of the "pellucid mind" was that it did not sacrifice complexity to achieve clarity, that it could take more into account without clouding the view. The illumination of which he speaks combines breadth of vision and depth of understanding with clarity of perception. It should not be confused with the intuitive perception sometimes associated with women's ways of knowing. It shares the holistic quality of intuition but grows out of the assimilation of numerous technical details rather than somehow circumventing them. For Whewell and his contemporaries, Somerville was not just an expositor or popularizer of science, not another descriptive nature writer who expounded on the limitless beauty of the landscape, but was one of the *illuminati*, one of the enlightened few who grasped the intricacies of science and higher mathematics with a certainty that allowed her to pass the torch of knowledge to others. Whewell described the process in this way:

> One idea after another, of those which constitute the basis of science, becomes distinct, first in the minds of discoverers, then in the minds of

all cultured men, till a general clearness of thought illuminates the land; and thus the torch of knowledge is handed forwards, thousands upon thousands lighting their lamps as it passes on; while still from time to time some new Prometheus catches a fresh light from heaven, to spread about among men in a like manner. (1837b: 27)

The image of knowledge being radiated outward from those who first perceive it reflects a dynamic view of the process by which new insights translate into general enlightenment or "clarity of thought," as well as a belief that the illuminating power of new ideas in science is amplified as those ideas become more widely known and understood. This pattern, which might be schematized as idea-illumination-enlightenment, reflects the power ascribed to those who could "catch a fresh light from heaven" and pass it on. The image of the torch of knowledge being passed from one hand to another often appeared on the covers of Somerville's books and functions as a visible symbol of the role that she and her work played.

Whewell further develops the idea of the illuminating female intellect in another poem, this one modeled on Dryden's "Lines on Milton." He replaces Homer, Virgil, and Milton with Hypatia of Alexandria, Maria Agnesi of Italy, and Mary Somerville.

Three women in three different ages born,
Greece, Italy and England did adorn;
Rare as poetic minds of master flights,
Three only rose to science' loftiest heights. (68)

His replacement of three great male poets by three great women of science is rooted in more than the fortunate accident of parallels in their cultures of origin. He draws on a rich tradition that associates greatness in literature and in science with the sublime vision revealed through a mind that sees more and sees more clearly than others, that comprehends both the vast and the minute, and that synthesizes it all into an awe-inspiring, unified, and coherent view.

People responding to Somerville's work often used language and images drawn from a poetic tradition that originated with Milton and evolved through the praise of Newton. This tradition has been vividly delineated by Marjorie Hope Nicolson in *Newton Demands the Muse* (1946). Nicolson argues that the poets of the eighteenth century saw no exaggeration in Alexander Pope's assertion that

Nature and Nature's laws lay hid in night;
God said, "Let Newton be!" and all was light.

The light of the eighteenth-century poets fused Miltonian and Newtonian elements. True science in the mold of Newton combined and rehearsed two miracles: "the creation of light in nature" and "the creation of man's reason." (Nicolson 1946: 37) It served as a powerful source of poetic inspiration: "Through their study of Newton, the poets came to look upon nature with new eyes. . . . he had added new beauty because he added new truth." (Nicolson 1946: 19, 32) Poets who studied Newton found that the natural world seemed both richer and larger than it had before. They saw Newton as an illuminator and cosmic voyager: "a mind for ever / Voyaging through strange seas of thought alone." (Wordsworth, *The Prelude*, III: 62–63) The crucial distinction for Somerville's readers was that she cast them in the role of fellow voyager.

Many of her contemporaries described Somerville in terms drawn from this tradition, and Whewell was not the only one who responded with poetry. In his *Lines Suggested by the Third Meeting of the British Association for the Advancement of Science* (1834), William Sotheby questioned why Somerville, who was by then an acknowledged member of the scientific elite, had not attended the meeting, which had been held in June of 1833 at Cambridge. Like Whewell's, Sotheby's lines convey the qualities he perceived in her person and her writing. He addresses her as "High gifted Somerville"; as "thou, in whom we love alike to trace / The force of reason, and each female grace"; and as a "cultured mind, / Smoothing the path of knowledge to mankind." Sotheby's poem also reflects the recognition Somerville had achieved, the esteem in which she was held: "While Cambridge glorying in her Newton's fame, / Records with his, thy woman's honoured name."

Twenty years later, the Italian countess Caterina Bon-Brenzoni also used poetry to convey her response to Somerville. Bon-Brenzoni's poem "I Cieli," or "The Heavens" (1853), illustrates the kind of response Somerville provoked, the kind of figure she was perceived to be, and the extent to which she mastered what Nicolson has labeled the "aesthetics of the infinite," an imagination-expanding response to the vastness and majesty of the universe. (Nicolson 1959: 140)

The poet responds to Somerville with trembling, reverence, and wonder, "overcome by love." Yet Somerville speaks to her "soft and low with the angelic voice and the humility that wisdom teaches to those who are dearest to her." Bon-Brenzoni addresses Somerville as "Kindly guide," one who has led her from sun to sun and through constellations: "God's own Pavilions! Temples of light!" Through Somerville, the poet has perceived the immensity of creation, concluding, "The pangs are on the same scale as that space and those worlds, emotion and thought are without limit." The encounter is clearly a mind-expanding experience as well as an uplifting and humbling one, and Somerville emerges from Bon-Brenzoni's poem a kind of Virgil figure, or perhaps more appropriately, a combination of Virgil, Beatrice, Milton, and Newton. In yet another blending of scientific and literary traditions, Bon-Brenzoni portrays Somerville as both the cosmic voyager of Newtonian science and the gentle guide of epic poetry.

Though it illustrates well the poetic inspiration Somerville's writing could provide, there is one respect in which Bon-Brenzoni's response and the other poetic responses are misleading. They are much more effusive than Somerville's writing, even at its most intense and poetic. The novelist Maria Edgeworth, in a letter to Somerville on May 31, 1832, reflected on this restrained poetic quality.

> The great simplicity of your manner of writing, I may say of your *mind*, which appears in your writing, particularly suits the scientific sublime – which would be destroyed by what is commonly called fine writing. You trust sufficiently to the natural interest of your subject, to the importance of the facts, the beauty of the whole, and the adaptation of the means to the end, in every part of the immense whole. This reliance upon your reader's feeling with you, was to me very gratifying. (PR 204)

The simplicity of mind and presentation to which Edgeworth alludes are associated with another aspect of the aesthetic tradition in which Somerville worked and with which Whewell and others associated her – the ideal of the supreme intelligence or divine mind, which "conserves the principle of order in the midst of perplexity." (Herschel 1832: 541)

A Dialogue Between Poetry and Science

There is a sense in which Somerville's writing can be understood as a dialogue between poetry and science. There are relatively few direct quotations from poetry in Somerville's scientific writing; the poetic element derives from the themes she uses and the scenes she sets. Still, her writing can be read as an implicit dialogue between scientific knowledge and the perspectives offered by poetry. This dialogue is carried on so subtly that it is often difficult to discern in her writing; what is going on is clearer in the notebooks she used to collect material for her books.[1]

---

[1] These notebooks are not very consistently or systematically organized, and it is sometimes unclear which of her publications they relate to. They are filled with extracts from scientific papers she has read, paraphrased accounts of new developments, factual statements, summaries, and observations as well as aphorisms, anecdotes, and poetry that provides perspectives on science. (Somerville Collection: Folder MSSW-5 Dep. c. 352)

Sometimes the material in the notebook is attributed to an author and publication; frequently, it is unidentified. The topics range from "Faraday on the Electric Telegraph" and "Researches on Light by Mr. A. J. Angström" to "Specific Heat" and "E. J. Cooper on the Perihelion and Nodes of the Planets." Sometimes the material takes the form of unattributed quotes, such as the following: "We must never forget that it is principles, and not phenomena; – laws not insulated and independent facts which are the objects of inquiry to the natural philosopher." And, "The ultimate object of all Science is to improve the character and condition of the human race." The notebooks constitute a kind of varied, wide-ranging, and very loosely organized conversation about science. This collection of material is synthesized, reworked, reorganized, and reused in a complex way. A part of a paraphrased passage will appear in one book, another part of the same passage in another book.

Some of the most interesting transformations have to do with poetry, which is directly quoted more often in the notebooks than in the books Somerville published. In one of the notebooks, the left-hand page contains a summary of a statement from William Herschel, indicating the number of stars that passed across the view through his telescope in an hour. Below this passage is a quotation from Byron concerning human insignificance in the grand scheme of nature. On the opposite page, there are extensive quotations from Byron's *Cain* (1905), which refer to the "beautiful, unnumbered, and endearing" suns "Not dazzling, and yet drawing us to them." Although the poetry itself does not, the images and sentiment conveyed by the poetry find their way into Somerville's text. Science, poetry, and philosophy are interwoven into a coherent and inspiring account. In note form, the account seems chaotic, even schizophrenic; in finished form, it is enormously rich.

An example of this kind of dialogue or interweaving occurs in *On the Connexion of the Physical Sciences*, where this summary of William Herschel's observations appears:

> Great as the number of comets appears to be, it is absolutely nothing when compared with the multitude of the fixed stars. About 2000 only are visible to the naked eye; but when we view the heavens with a telescope, their number seems to be limited only by the imperfection of the instrument. In one hour Sir William Herschel estimated that 50,000 stars passed through the field of his telescope, in a zone of the heavens 2° in breadth. This, however, was stated as an instance of extraordinary crowding; but, on an average, the whole expanse of the heavens must exhibit about a hundred millions of fixed stars within the reach of telescopic vision. (Conn 361)

One might be tempted to think that the poetry had been abandoned, since twenty pages of detailed scientific summary follow this passage, as Somerville surveys the cataloguing of stars; binary, multiple, and nebular systems; the laws of motion as they apply to the movements of the stars; and the distances of the stars from earth and their distribution in the heavens. Before she moves into a discussion of meteorites and shooting stars, she pauses for an internal summary that transforms Byron's poetic sentiments and perspective, blends them with Herschel's observations, and presents an expertise-based portrait of the stars as seen not from the earth but from the heavens:

> So numerous are the objects which meet our view in the heavens, that we cannot imagine a part of space where some light would not strike the eye; – innumerable stars, thousands of double and multiple systems, clusters in one blaze with their tens of thousands of stars, and the nebulæ amazing us by the strangeness of their forms and the incomprehensibility of their nature, till at last, from the limit of our senses, even these thin and airy phantoms vanish in the distance. (Conn 381)

She then uses this view from the heavens, what I will call here the cosmic platform, to engage the reader's imagination and to provide an even more expansive view that includes not only the heavenly bodies we can see, but also those we cannot.

> If such remote bodies shone by reflected light, we should be unconscious of their existence. Each star must then be a sun, and may be presumed to have its system of planets, satellites, and comets, like our own; and, for aught we know, myriads of bodies may be wandering in space unseen by us, of whose nature we can form no idea, and still less of the part they perform in the economy of the universe. (Conn 381)

This passage, like many others in Somerville's works, reveals not only her powerful intellect, but her imaginative and illuminated soul.

The Scientific Sublime as the Fusing of Traditions

Somerville's writing, then, evokes the scientific sublime, the capacity of the vision of nature revealed through science to summon forth the same sense of majesty and power that human beings feel in the presence of God.[2] Somerville combines vision on a cosmic scale with a restrained poetic quality, even as she presents the substance of abstruse and recondite science. The scientific sublime also raises the reader

---

[2] In *Mountain Gloom and Mountain Glory* (1959), Nicolson defines the essence of the sublime as follows: "The manifestations of God's majesty and power in Nature must evoke in sensitive minds some degree of the awe they feel for God Himself, which is the essence of the Sublime experience." (282)

into a realm beyond ordinary human experience, as demonstrated in a discussion of the propagation of sound. Following an extended analogy comparing light, heat, and sound, Somerville moves the reader smoothly and rapidly from earthly phenomena to an empyrean view, a cosmic platform:

> above the surface of the earth, [where] the noise of the tempest ceases and the thunder is heard no more in those boundless regions, where the heavenly bodies accomplish their periods in eternal and sublime silence. (Mech lviii)

This platform strikingly resembles the regions of calmness and serenity above "this dim spot / which men call earth" to which Whewell refers in his review. (1834: 65)

Like the poets on whom she drew, Somerville was profoundly affected by the view of nature revealed through the telescope, the microscope, and other tools of science. She was also extremely skillful in recreating the experience of the scientific sublime for her readers. But, unlike the scientific poets of the eighteenth century or the prolific nature writers of the nineteenth century, Somerville was able to convey the detailed substance of science in an authoritative manner even as she provided the reader with a great deal of pleasure. Close examination of her works shows that only a small portion of each was devoted to creating the framework and developing the themes that evoked the scientific sublime. In the bulk of her work, she offered technically precise and highly detailed accounts of the substance of science and mathematics. She presented science as both "exact calculation" and "elevated meditation." She took the poetic traditions established by Milton, the eighteenth-century poets, and the early nineteenth-century Romantic poets and transformed them for scientific prose. In the process, she created a powerful and persuasive rhetoric for science, which relied on new ways of seeing and responding to the natural world. She also established herself as one of most eminent scientists of her day.

Somerville rose to eminence by teaching people to use science and a variety of existing aesthetic traditions to see and respond to the natural world. In an era of rapid scientific and social change, she was able – through her person and her writings – to represent science as a progressive and enlightened enterprise that was compatible with, indeed supportive of, moral, religious, and aesthetic traditions. She marshaled the resources of the poets and the leading lights of science in the effort, and she showed how the illumination of science could be transformed into enlightenment for humankind; human imagination empowered through science could cross magnitudes of space and time; language could be used to help imagination construct what could not be directly observed.

Countless examples demonstrate that "early Victorian scientists . . . had a love for what lay outside the experience of the ordinary man" (Schweber 1981: 18), and much of the appeal of science in the Victorian era lay in its capacity to enable human beings to get beyond what they could know through ordinary experience. The power of science was perhaps best symbolized in the predictive power of physical astronomy and geology, where accurate and powerful inferences about past and future could be made based on very little directly observed data. Although Somerville recognized that both language and imagination sometimes fail in their attempt to describe the natural world, her own writing shows the extent to which language can be used to push the frontiers of imagination. Her work and her reader's responses to that work reveal the ways in which science, like poetry, can be conceived as a heightened form of perception and expression. Writing in an era before electronic media, when views on a cosmic scale could be constructed only in words and reconstructed only in imagination, Somerville presented a vision of the universe whose magnitude and immensity gave readers pleasure, expanded their minds, and enlarged their conception of the universe. She employed both science and literary imagination in the effort and blurred the boundaries of gender and genre in the process.

1

Head Among the Stars, Feet Firm Upon the Earth

The Problem of Categorizing Mary Somerville

Mrs. Somerville is the lady who, Laplace says, is the only woman who understands his works. She draws beautifully, and while her head is up among the stars, her feet are firm upon the earth.

— Maria Edgeworth to Miss Ruxton,
January 17, 1822

The innovative is, by definition, hard to categorize.

— Clifford Geertz,
"Blurred Genres: The Reconfiguration of Social Thought"

Mary Somerville was an eminent scientist. She achieved an international reputation that established her as both the leading woman of science in Great Britain during the nineteenth century and as one of that century's most celebrated intellectual women. (Patterson 1983)

Somerville's greatest scholarly strength was in mathematics, which she mastered at a very high level, but her expertise extended throughout the established and emerging physical and life sciences. She was the first woman to publish experimental results in the *Philosophical Transactions of the Royal Society* as well as the first – and only – woman to have her bust placed in the great hall of the Royal Society.[1] Hers was the first name on John Stuart Mill's petition to obtain the vote for women in Great Britain. (PR 345) When the founders of Oxford University's first nondenominational women's college sought a name to exemplify ideals of high intellectual achievement for women, "Somerville" seemed an obvious choice. (Adams 1996)

In a letter written in 1829, David Brewster pronounced her "certainly the most extraordinary woman in Europe – a Mathematician of the very first rank." The *Morning Post*, in an obituary notice published on December 2, 1872, judged her "one of the most distinguished astronomers and philosophers of the day" and asserted that, while there might be debate as to who was the leading man of science in the mid–nineteenth century, there was no question that Mary Somerville was "the queen of science." On March 25, 1874, *The Kelso Mail* called her "the most remarkable woman of her time." On November 25, 1922, the *Times Educational Supplement* asserted that her works "were perhaps the first scientific works to be accepted and weighed on equal terms, by the specialists of the age.... It is doubtful if any scholar of her age received greater recognition." In 1941, A. W. Richeson, writing in *Scripta Mathematica*, concluded: "She is undoubtedly the greatest woman scientist that the English have ever produced." (13)

As a woman scientist – and even more as an eminent woman scientist – Mary Somerville defies the categories for thinking about the intersections of science, women, and history. She defied the categories as they existed in her own time, and she defies the categories that we use today, though in more subtle and somewhat different ways. Her career in and contributions to science constitute an innovative, creative response both to the constraints within which she worked and to needs of the scientific community of her day.

The value attached to her writing and the eminence she achieved are accounted for by a complex of factors, a mixture of resistance and acceptance in which she duplicated the principle of counterpoise that she captured in her portrayal of nature and embodied in her person. A

---

[1] For a discussion of women and the Royal Society, see Rose (1994) and Mason (1992, 1995, 1996). None of these sources treats Somerville in detail, but all provide useful background.

full appreciation of her accomplishments can be achieved only within a framework that draws on both the history of science and literary history and through a willingness to abandon many conventional notions about the history of science and the role of women in it.

The Problem of Naming

The problem of naming – of finding appropriate language or categories in which to account for feminine accomplishments – is well recognized in contemporary scholarship on women, gender, and science. (Gates and Shteir 1997: 19; Keller 1995: 29; Lightman 1997: 63; Rose 1994: 3) The extent to which Somerville's defiance of the categories of her own day provoked creative responses is exemplified in a review of one of her books. The book was *On the Connexion of the Physical Sciences*, which was published in 1834. The reviewer was William Whewell, Master of Trinity College, Cambridge, and one of the leading men of science of the nineteenth century. The place was the *Quarterly Review*, a major periodical that covered a wide range of topics and also served as a forum for many of the advanced scientific debates of the day. Although Whewell is explicitly dealing with *Connexion*, he also had her first book, *Mechanism of the Heavens* (1831), in mind. *Mechanism* was a translation, elucidation, and interpretation of the *Mécanique Céleste* of Pierre Simon Laplace. *Connexion* was an extended literature review; it offered a comprehensive, state-of-the-art survey of knowledge in all areas of the physical sciences.

   Whewell clearly wrestled with the problem of reconciling the quality of Somerville's mind and accomplishments with conventional prejudices about the intellectual capabilities of women. Like his contemporaries, he believed there was "a sex in minds." (Whewell 1834: 65) But, unlike most other commentators, including some of Somerville's most faithful supporters, he was not satisfied simply to consider her an exception to the rule of female limitations in higher intellectual pursuits. He was so struck by Somerville as a case where "all our prejudices against such female acquirements vanish" (65) that he attempted to create a new category and quality of mind to accommodate her. The category is a very small and elite set of "eminent female mathematicians" of whom there have been only two others "in the history of the world." (66) The quality of mind is the "peculiar illumination," which adds to "the great merit of being profound . . . the great excellence of being clear." (65) It is peculiar in the sense of being special to, perhaps even distinctively associated with, female minds. One of the most notable aspects of this illumination is that it is not an inferior or reflected version of the male intellect but is in some respects superior to the male philosophical mind.

The esteem in which Whewell holds Somerville and her work emerges clearly in the review. He describes *Connexion* as an "able and *masterly* [his italics] . . . exposition of the present state of the leading branches of the physical sciences" (56) and Somerville herself as a "person of real science" who has "condescended" to write for "the wider public." (57) He praises her work for its currency, conciseness, and clarity; its scrupulousness in distinguishing conjectures from assertions; and for its "noble object . . . of establishing connections among developments in various areas of science." (56)

Despite his admiration for Somerville, it is apparent that Whewell works within an essentialist framework and shares many common notions about gender and intellectual activity. He describes women as limited in their ability to grasp and to reason about intellectual matters, largely because their emotions and intuitions are much more fully developed than their intellects. "With them," he says, referring to women, "action is the result of feeling, thought of seeing. . . . The heart goes on with its own concerns, asking no counsel of the head. . . . the working of the head (if it does work) is not impeded by its having to solve questions of casuistry for the heart." (65)

Given the gender stereotypes operative in the nineteenth century (Bell and Yalom 1990; Benjamin 1991a and b; David 1987; Hale 1840; Helsinger et al. 1983; Hollis 1979; Jordanova 1997; Kolhstedt 1995; Kohlstedt and Longino 1997; Mellor 1993; Outram 1987; Phillips 1990; Russett 1989; Shepherd 1993), none of this is surprising. What is surprising is that Whewell does not oppose this archetypal woman of well-developed heart and inoperative intellect to what Genevieve Lloyd (1984) has called "the man of reason," but rather to a man "lost in a cloud of words" and subject to "inextricable confusion – an endless seesaw of demand and evasion" because his heart and head interfere with each other. Referring to "the man," he says, "The heart and the head are in perpetual negotiation, trying in vain to bring about a treaty of alliance, offensive and defensive."

Whewell's distinctively illuminated female mind is defined in relation to this "man . . . lost in a cloud of words." The female mind is elevated above the conflicts that perplex the male mind and is thus able to theorize "In regions mild, of calm and serene air, / Above the smoke and stir of this dim spot / Which men call earth." (65) Referring to the male tendency to be mystified by conflict and complexity, Whewell asserts:

Women never do this: what they understand, they understand clearly; what they see at all, they see in sunshine. . . . from the peculiar mental character to which we have referred, it follows, that when women are philosophers, they are likely to be lucid ones; that when they extend the range of their speculative views, there will be a peculiar illumination thrown over the prospect. If they attain to the merit of being profound, they will add to this the great excellence of being clear. (65–66)

Although much of Whewell's reasoning is confused, it is nonetheless instructive to watch him struggle to develop a category that will accommodate Somerville and resolve the tension between his conventional ideas about women and the qualities he perceives in her. The resulting concept of the distinctively illuminated female mind is no less gendered than the stereotypes with which he starts, but it does create a space, if a small and highly specialized one, for women in the realm of the highest intellectual pursuits. Dealing with an exceptional woman forces him to articulate and, more importantly, rethink and revise his assumptions about women, gender, and science. One of the aims of this book is to replicate and amplify the conceptual expansion and revision begun in Whewell's analysis – to account more fully for Somerville's eminence and to rethink her relationship to the standard categories for conceptualizing women and science.

Rethinking Somerville

Of all the women – and there were many – who defied gender categories to become scientists, Somerville is probably one of the least forgotten. She has been treated extensively in a book-length biography (Patterson 1983), been the subject of numerous brief biographies (Adams 1884; Basalla 1963; Hamer 1887; Johnson 1994; Neeley 1992; Osen 1974; Patterson 1969, 1974, 1975; Perl 1978; Richeson 1941; Tabor 1933; Toth and Toth 1978), appeared routinely in comprehensive surveys (Alic 1986; Mozans 1913; Ogilvie 1986), and been included in many analytical historical works dealing with women, gender, and science (Benjamin 1991a and b; Cannon 1978; Helsinger 1983; Hollis 1979; Phillips 1990; Rose 1994; Rossiter 1982). She is a staple of the historical genre Sylvana Tomaselli (1988) calls "collecting women" and Hilary Rose (1994) calls "rescuing Hypatia's sisters." The question might well be asked, why is Somerville worthy of further study?

The first reason is that the existing scholarship, while quite substantial, provides only a partial view of Somerville's work and the factors that contributed to her success. The majority of the brief biographies and analytical works treat Somerville's scientific writings superficially and often mischaracterize her work on the assumption that it bears more resemblance than it actually does to the work of other women in science. Elizabeth Patterson's *Mary Somerville and the Cultivation of Science 1815–1840* (1983) does an admirable job of using Somerville's voluminous correspondence to piece together a detailed chronology of the first half of Somerville's career, establishes the eminence Somerville achieved, and offers an intricate portrayal of Somerville's relationship to other members of the scientific community, but it is limited in a number of significant ways.

The first limitation is that Patterson devotes only a little over a page to the last thirty-two years of Somerville's life and gives only a paragraph to each of Somerville's last two books, one of which was very popular. Another limitation is that, although Patterson acknowledges Somerville's skill as a writer, she does not provide any detailed analysis of the techniques that gave Somerville's writing its significance, clarity, and appeal. Patterson's work does not even hint at the power of Somerville's imagination or her rhetorical skill. Additionally, since much of the most interesting work on women, gender, and science was published after Patterson's project was well under way, she was not able to take advantage of that work and consequently did not give adequate consideration to Somerville's place within the larger picture created through scholarly work in these areas. Finally, Patterson works for the most part uncritically within a framework that presumes that making discoveries is the only way to earn a place in the history of science. Although the phrase "cultivation of science" appears in Patterson's title, she does not discuss cultivation as a distinct form of activity. None of this should obscure the enormous scholarly contribution that Patterson's biography represents. My work would have been impossible without it, but it leaves a great deal about Mary Somerville unsaid.

The second reason for looking at Somerville comprehensively and in more depth is the eminence she achieved. The term "eminence" is intended to reflect the recognition she received, and is not intended to establish any kind of hierarchy of women scientists. Such ranking would not have been appealing to her, and is not appealing to me. The intent, rather, is to give greater attention to what Hilary Rose has called "women at the apex of the prestige system of science." (1994: xii) There is ample evidence provided by Patterson and by the evaluations of Somerville's contemporaries to support the argument that Somerville occupied such a position, and it is worth exploring her career from this angle because most history of science has operated at least implicitly within an assumption that the apex of the prestige of system of science was, by definition, a place where there were no women.

A third reason for studying Somerville is that her career and work allow us to see both science and gender as powerful, multifarious, fluid, and historically variable entities. Scholars working in a variety of areas have highlighted the ways in which boundaries, definitions, dualistic thinking, and hierarchies in science have shifted drastically over time.[2] Feminist studies of women and gender (gender taken here to distinguish the sociocultural from the biological) have revealed far more layers and diversity in the culture of science than had been recognized by traditional historians of science. In the words of Kohlstedt and Longino, "The sciences are diverse, constituted by a multiplicity of practice. Further, the world *science* itself is used in various ways." (1997: 13) As many of the responses to Somerville analyzed later in the book demonstrate, Somerville constitutes what Mary Poovey (1991) has called a "border case" that is problematic because it "marks the limits of ideological certainty." (12) The attempts of commentators to resolve the contradictions Somerville presented provide a window into the dynamics of gender along with an opportunity to observe changes in gendered thinking during her relatively long lifetime. Whewell's commentary alone provides an excellent example of what Ludmilla Jordanova has called "the multiple relationships between masculine and feminine" and the capacity of gender "to represent so many ideas." (1993: 474)

A fourth reason for reassessing Somerville is that, despite the recovery of women's historical accomplishments in science, the insights of gender studies, and the recognition of the broad cultural dimensions of science, women's history remains almost entirely unintegrated into larger historical narratives in science.[3] A case like Somerville's clarifies the limitations of our current paradigm and vocabulary and demonstrates the plausibility of a much more satisfying narrative of women's participation in science.

The narrative I propose is derived from the work of historians seeking to account for women in the history of art. (Parker and Pollack 1981) That history varies in some significant ways from the history of women in science, so the fit is not exact, but there are enough similarities that their argument can be used effectively for women in the history of science. In brief form, the approach, which I will call "integrative," is as follows:

---

[2] The last twenty years have produced a very rich literature aimed at getting beyond monolithic, ahistorical conceptions of science and illuminating the range, diversity, and historical evolution of concepts and activities related to science. A few of the most notable studies include Cannon (1978), Crosland (1976), Dale (1989), Golinski (1992), Jardine, Secord, and Spary (1996), Jordanova (1993), Laudan (1993), Morrell and Thackray (1981), Shinn and Whitley (1985), and Yeo (1987).

[3] The failure to integrate women's accomplishments into larger narratives of the history of science has been noted by many scholars, including Keller (1995), Kohlstedt (1995), and Kohlstedt and Longino (1997).

There have always been women scientists. Because they have operated within a predominantly male culture, they have occupied spaces different from those occupied by men. This necessity has contributed to, indeed forced, much of the creativity women have exhibited in their scientific work. It has also meant that there has been a great deal of variety in the ways that women have defined, used, and .attempted to enlarge the spaces made available to them within science. Although women have been noticeably absent from standard historical accounts of science, they have been very much present in lived history both as the negative feminine against which the positive qualities of masculinized professional science were defined and as active participants and contributors to science within the positions they were able to negotiate.

This narrative accommodates and reflects many of the major insights of scholarship on women, gender, and science.[4] It provides a view of women in the history of science that both integrates them into the larger picture and differentiates their experience and contributions. One of the greatest strengths of this approach is that it gets outside the vocabulary of "marginality," which is particularly distorting in discussions of the relatively few women who have achieved eminence. Somerville's own career provides a fully realized example of the explanatory power an integrative model can have. Retelling Somerville's story, then, becomes a point of entry both for highlighting the particularities of Somerville's career and for rethinking the story of women's participation in science.

I focus this retelling by emphasizing the way Somerville's contemporaries perceived her and by exploring the meaning and manifestations of the concept of the distinctively illuminated female mind. The use of the phrase "the female mind" is, of course, somewhat ironic. Modern scholarship on gender has dismissed the essentialist notion of *a* or *the* female mind, just as it has almost entirely dismissed the notion of *a* or *the* male mind. And I do not believe personally that such an entity as the female mind exists. But the fact remains that such concepts have been important historically and were especially so during Somerville's own lifetime. As a particular description of Somerville – and I believe that is what it really was – the distinctively illuminated

---

[4] The notion that there have always been women scientists is amply supported, not only by the comprehensive historical treatments of Mozans (1913), Alic (1986), and Ogilvie (1986), but also by the work of Abir-Am and Outram (1987), Benjamin (1991), Phillips (1990), and Rossiter (1982). The clearest evidence of the ways that the feminine has acted as the negative against which a masculinist science is defined is provided by Schiebinger (1989), Rose (1994), and Gates and Shteir (1997). The work of David (1987) and Poovey (1991) demonstrates that the patterns of acceptance mixed with resistance and uneven deployment of gendered thinking are larger contexts within which the case of women in science can be understood.

female mind becomes a useful focus for exploring the qualities of Somerville's person and work that provoked such strong responses from Whewell and others.

Who Was Mary Somerville?

This apparently simple question has no simple answer, largely because, as mentioned, Somerville's accomplishments and influence are not adequately accounted for by the categories and patterns ordinarily used to organize literary history, the history of science, or even the history of women in science. Though her scientific writings draw extensively on a number of rich literary and aesthetic traditions, they do not exemplify any clearly recognized literary genre. Though she played a prominent role in nineteenth-century science, her story cannot be told as a life structured around significant scientific discoveries. One way to begin is to follow the lead of Maria Edgeworth and define Somerville in several ways as "the woman who . . ." – who possessed particular abilities, expertise, and personal attributes; who had certain experiences and accomplishments; and who viewed the world and was viewed by others in a particular way. This is perhaps the best place to start because there is a remarkable consistency in the descriptions of Somerville provided by her contemporaries.

Maria Edgeworth's description, as quoted at the beginning of this chapter, is a typical though particularly well expressed example. This characterization was repeated by several others, all of whom emphasized Somerville's status as the only woman who fully comprehended the work of Laplace, who was himself viewed as the greatest scientist of his day and second to Newton alone.[5] Somerville's mastery of Laplace carried even more weight because the great man himself had recognized her knowledge and ability, in effect conferring the mantle of his approval upon her. Thus, the first element of the characterization reflected the transcendent intellect readers perceived in Somerville's work and her elite status as one of the very few women who had risen "to science' loftiest heights . . . one of the brightest ornaments of England." (Whewell 1834: 68) This intellectual achievement

---

[5] Edgeworth had another version in which Somerville was "the only woman who could correct his works." In yet another version, Laplace himself was reported to have said that there were only two women besides Mary Somerville who understood his work – a Miss Fairfax and a Mrs. Greig – all three of whom were, of course, united in Mary Fairfax Greig Somerville. The version quoted here is taken from a copy of a letter that is included in the Somerville Collection, Folder MSE-1 in Dep.c.369.

was the foundation of her greatness, the thing without which her other qualities and abilities would have gone unnoticed.

The second element of the characterization – "she draws beautifully" – conveys meaning on several levels. At the literal level, it refers both to Somerville's highly developed aesthetic sense and her artistic ability. She studied painting under the noted landscape artist Alexander Nasmyth, gained recognition for her talent, and enjoyed painting, mostly landscapes, throughout her life. The self-portrait that serves as the frontispiece for this book provides evidence of he artistic ability. But her painterly skills of observation and representation and her strong aesthetic sensibilities were perhaps most fully exercised in her writing, which is permeated with her perception of "the exquisite loveliness of the visible world." (PR 348)

When William Whewell attempted to describe the effect of Somerville's writing, he used an analogy between the verbal and visual presentation of information.

> The office of language is to produce a picture in the mind; and it may happen . . . that we are struck by the profound thought and unity displayed in the colouring, while there is hardly a single object outlined with any tolerable fidelity and distinctness. The long-drawn vista, the level sunbeams, the shining ocean, spreading among ships and palaces, woods and mountains, may make the painting offer to the eye a noble expanse magnificently occupied; while even in the foreground, we cannot distinguish whether it is a broken column or a sleeping shepherd which lies on the earth. . . . In like manner, language may be so employed that it shall present to us science as an extensive and splendid prospect, in which we see the relative positions and bearings of many parts. (Whewell 1834: 55)

Whewell's description highlights one of the hallmarks of Somerville's writing – its striking visual quality and capacity to paint pictures with words in order to allow readers to visualize complex processes and vast expanses in nature. The phrases "profound thought and unity," "noble expanse magnificently occupied," and "extensive and splendid prospect" all reflect Somerville's capacity to evoke the aesthetic perception of wholeness, vastness, and harmony associated with the sublime. The analogy between the verbal and visual presentation of information is very helpful in an understanding of Somerville's work, but it is important to recognize that she was much more poet and landscape artist than storyteller or cartographer in her use of words.

At a symbolic level, Somerville's ability "to draw beautifully" exemplified suitable feminine acquirements, her possession of womanly traits and conformity to feminine norms. In his poetic comparison of Somerville with Hypatia of Alexandria and Maria Agnesi, Whewell gave Somerville the place of honor Dryden had given to Milton. In Whewell's view, Somerville shared Hypatia's and Agnesi's expertise and transcendent intellect as well as their ability both to grasp and to teach complex concepts. She also shared their virtue, but she surpassed them because of the happiness she achieved through her conformity to the domestic ideals established for women. Hypatia had been literally torn to pieces by an envious and brutal mob; Agnesi eventually abandoned mathematics and science and became a nun. Neither had ever married. But Somerville followed a different path, the path of marriage, family, and domestic duty.

> Equal to these, the third, and happier far,
> Cheerful though wise, though learned popular,
> Liked by the many, valued by the few,
> Instructs the world, yet dubbed by none a Blue.
>   (Whewell 1834: 68)

For Whewell and many others, Mary Somerville was a model of the happiness to be achieved through the pursuit of early Victorian notions of ideal womanhood and the rejection of feminism.[6] This ideal did not absolutely forbid the acquisition of scientific expertise, but it made the serious pursuit of science very difficult.

Probably the most commonly made statement about Mary Somerville was that she was a thoroughly feminine woman who had succeeded in demonstrating the compatibility of the highest in intellectual pursuits and an exemplary performance as a wife and mother. Given the times in which she lived, Somerville's conventional femininity was at least as important as her intellectual accomplishments in establishing and maintaining her reputation. The world into which she was born was likely to ridicule intellectual women but not much inclined to fear them. The world in which Somerville lived her adult life became increasingly preoccupied with the role of women and increasingly fearful that they would fail to acknowledge the claims that nature and convention made upon them by choosing not to become wives and mothers or by refusing to accept a subordinate place in society. Despite her superior intellectual accomplishments, Mary Somerville seemed firmly grounded in feminine ideals of self-effacing womanhood. Devoted mother, dutiful daughter, adoring wife – these were roles that she performed with grace and ease and that she continued to relish even after she became the famous translator and interpreter of Laplace.

Her high level of accomplishment in science and mathematics often seemed to be the only thing that set her apart from other women. As

---

[6] For a discussion of Victorian ideals of womanhood see David (1987) and Poovey (1991).

one reviewer of Somerville's autobiography put it, her life offered "a picture of the most attractive character, of a woman to whom you can deny nothing – of one who, with all her transcendent talent, could be just on a level with ourselves."[7] The author of the review was not sorry to have missed Madame de Staël, Margaret Fuller, or the mother of the Gracchi but would have liked to know Mary Somerville.

This last comment leads to the third element of Edgeworth's characterization – Somerville's ability to bring together the divergent and sometimes conflicting realms symbolized by having "her head up among the stars" and "her feet firm upon the earth." This pattern suggests that her greatness lay not in any particular attribute or set of attributes, but in the way she brought together apparently irreconcilable elements – realms and qualities presumed either to be at odds or to be hopelessly separated – and created something richer and stronger in the process. This ability is demonstrated on many occasions in Somerville's work as it weaves together poetry and science, imagination and mathematics, religion and science, the world of the telescope and the world of the microscope, the cosmic and the everyday.

These three elements – transcendent intellect, womanly behavior, and the capacity to reconcile apparent oppositions – were repeated in nearly all descriptions of Somerville, though the element emphasized varied from one description to another. When David Brewster described Somerville to John David Forbes in 1829, he mentioned traits very similar to the three mentioned by Edgeworth: Somerville's mastery of mathematics and physical science, combined "with the gentleness of a woman" and "all the simplicity of a child."[8] Associated with the transcendent perspective captured in the phrase "head among the stars," there was also a quality in Somerville's writing and her person that was at once accessible, feminine, and profoundly down to earth. In an era when it was widely assumed that women either lacked the intellectual capability to master higher mathematics and science or would lose their femininity if they did, Mary Somerville conformed to all important norms of womanly behavior at the same time that she challenged many assumptions about women and science.

Although the analysis of the factors that contributed to Somerville's development and eventual success is very complex, her life story is straightforward and, especially in the early years, typical for a woman of her era, country, and social class. It provides a factual outline that

---

[7] *The Literary World*, December 19, 1873, pp. 385–388.
[8] David Brewster to John David Forbes, September 11, 1829 (J. D. Forbes Papers, St. Andrews University).

will be refined, expanded, and interpreted as subsequent chapters examine in more detail the events of Somerville's life and the character of her work; it will serve as a rough sketch that should eventually become itself "an expanse magnificently occupied."

The Life Story of Mary Fairfax Greig Somerville

Mary Somerville was born Mary Fairfax on December 26, 1780, in the manse at Jedburgh, Scotland, a town situated on the River Jed, not far from the English border. Her father, William George Fairfax, an officer in the English navy, had just left for a long period at sea. Mary's mother, Margaret Charters Fairfax, had accompanied him to London to see him off. Margaret had just arrived at the home of her sister Martha Charters Somerville and her husband, Rev. Dr. Thomas Somerville, minister of Jedburgh, when Mary was born. Because Margaret Fairfax was ill after the birth, the infant Mary was nursed by her Aunt Martha, who many years later became also her mother-in-law. More than thirteen years later, Thomas Somerville would also play a decisive role in young Mary's intellectual development. His son William would eventually play an even more important role as Somerville's second husband. Fortunate coincidences and connections were a significant factor throughout Somerville's life and career.

Mary spent most of her early childhood in Burntisland, a small seaport town across the Firth of Forth from Edinburgh, living in a house owned by her grandfather, Samuel Charters. She had an older brother, Samuel, who was away at school in Edinburgh much of the time (he lived with his Grandfather Charters). Her younger brother, Henry, was her only other sibling who survived to adulthood. For most of Mary's childhood, her father was away at sea, and Mary Fairfax lived a quiet life with her mother, Margaret, who was an indulgent, though "old-school," parent. Margaret Fairfax taught her daughter to read the Bible and saw to it that she learned her catechism, but did not provide any other formal education. Somerville portrays her mother as characterized by a mixture of courage and irrational fear. She was, for example, a woman who could organize her belongings in case a rapidly moving fire in the next building forced evacuation, and then sit down calmly and eat her breakfast while she waited to see whether evacuation would actually be necessary. Yet she was terrified beyond reason of making the short water crossing of the Firth of Forth from Burntisland to Edinburgh, and would not cross "except in the boat of a certain skipper who had served in the navy and lost a hand." (PR 48–49)

The solitude and freedom of life at Burntisland were punctuated by the periodic return of Mary's father, Lieutenant (later Admiral Sir) William Fairfax, a man of courage and presence of mind who fought in the war for American independence and distinguished himself in battle. He received honors but no financial rewards for his accomplishments, and Mary Fairfax grew up in what might be described as genteel poverty. When she was eight or nine, Admiral Fairfax returned home to find that she could not write or do simple sums. His own attempt to further her education by having her read from *The Spectator* and Hume's *History of England* was unsuccessful.

Perhaps out of a sense of duty, perhaps out of desperation, William Fairfax sent his daughter away at the age of ten or so to Miss Primrose's School in Musselburgh, where she reported herself "utterly wretched." (PR 21) Miss Primrose's methods of teaching strike the modern observer as having been explicitly calculated to inspire a distaste for learning. The primary mode of instruction was to have the girls copy and memorize pages from Johnson's dictionary. They undertook their lessons "enclosed with stiff stays with a steel busk in front," shoulder blades drawn back until they met, and "a steel rod, with a semicircle under the chin." (PR 22) Although she studied handwriting and the basics of English and French grammar, Somerville learned very little and "was reproached for having cost so much money in vain." (PR 24)

Young Mary Fairfax returned to Burntisland after her year of boarding school with a growing desire to educate herself, though it took her some time to establish the direction she wanted to take and to acquire the resources she needed. During a typical year, she spent the summers in Burntisland and the winters in Edinburgh, where she took music and painting lessons, as well as instruction in handwriting and arithmetic. She learned to read French, Latin, and Greek, largely on her own, and spent her summers in Burntisland honing her skills and expanding upon the knowledge she had acquired during the winters in Edinburgh. At the age of fourteen or fifteen, she was exposed to algebra through a magazine for ladies; it seemed entirely mysterious to her. Though she made little immediate progress, she persisted in her desire to learn more about mathematics.

As a young woman, she led an active social life. She attended dancing school; enjoyed breakfast, tea, and supper parties; met interesting people; and went to numerous balls and to the theater. Though basically shy, she thrived in the social atmosphere of Edinburgh. At home, she did her share of household duties and continued to develop womanly skills in music, painting, needlework, and pastry making. Increasingly ambitious to prove the abilities of women, she pursued her other studies with diligence, but, at least as she saw it, without results.

In 1804, she married a cousin, Captain Samuel Greig of the Russian navy, whom she had met when he came to visit the family at Burntisland. Prior to the marriage, Greig had obtained a permanent post as Russian consul in London, where the couple lived until Greig died in 1807. They had two sons, the younger of whom died in 1814 at the age of nine. The other, Woronzow Greig, named after a distinguished colleague of Samuel Greig, became a barrister. He died in 1865, seven years before his mother.

After a period of widowhood in Burntisland and Edinburgh, Mary Greig married again in 1812, this time to another cousin, William Somerville, son of her aunt and uncle, Martha and Thomas Somerville. During her widowhood, Mary Greig had enjoyed for the first time the opportunity and financial resources to pursue her mathematical studies in earnest; her marriage with William Somerville offered her encouragement and an opportunity to expand both the range of her studies and her network of scientific and intellectual associations.

In the early years of their marriage, the Somervilles lived in Edinburgh, where William worked as head of the army medical department. In 1816, his job took them to London, where they lived in Hanover Square and became part of a social circle that included many of the leading literary and scientific figures of the day. Their daughter Margaret was born in 1813, their daughter Martha in 1815, and their daughter Mary Charlotte in 1817. A son died in infancy in early 1815.

A change of position in 1819 for William Somerville meant a move to less pleasant quarters at Chelsea Hospital, London, where the Somervilles still kept up an active social life, and where they became friends with Lady Noel Byron, her daughter Ada, Sir Richard Napier and his wife, and Maria Edgeworth. Despite the untimely death of their eldest daughter, Margaret, in 1823, Mary Somerville continued her mathematical and scientific studies through private study and through almost daily contact with leading figures in science and mathematics, both at home and abroad. Somerville made many of her European contacts during tours of Europe in 1817–1818, 1824, and 1832–33. During each trip, she and William renewed old friendships and made new ones.

Mary Somerville first became a scientific author in 1826, when she published a paper "On the Magnetizing Power of the More Refrangible Rays" in the *Philosophical Transactions of the Royal Society at London*. She published a second paper on a similar subject in the *Philosophical Transactions* in 1845. In early 1827, she received a request from Lord

Henry Brougham to write "an account" of Laplace's *Mécanique Céleste*, which was eventually published by John Murray in 1831 under the title *Mechanism of the Heavens*. It was especially well received at Cambridge, where it became an important resource in the effort to modernize mathematics instruction. From 1831 onward, she continued to publish books on science. Like Somerville's career as a whole, her books resist easy characterization; the brief descriptions provided here are supplemented by much more detailed discussions in later chapters.

Mary Somerville began work on her second book, *On the Connexion of the Physical Sciences*, immediately after *Mechanism* was published. *Connexion* came out in early 1834. Though still not a popularization in the usual sense, it was directed toward a much broader audience than *Mechanism* had been, and it was a commercial as well as a scholarly success. Much of *Connexion* was written in Paris. Although generally healthy, Somerville had suffered badly at Chelsea from headaches, she had worked very hard on her manuscripts and research, and the family had experienced serious financial problems. Paris offered a change of scene and a more affordable lifestyle, and was still close enough so that William Somerville could visit as his responsibilities at Chelsea allowed. In 1835, she received a government pension in recognition of her work. In that same year, she published an extended discussion of recent and historical developments in the understanding of comets in the *Quarterly Review*.

Thus, from 1816 until 1838, Somerville spent most of her time in London in the midst of a brilliant social and intellectual circle, with access to up-to-date information about new discoveries in science, establishing herself as a respected scientist and writer of scientific books. In 1838, William Somerville suffered a long and serious illness. As soon as he was well enough to travel, the family went to Rome in the hope that a warmer climate would restore his health. That move to Italy turned out to be permanent.

At Rome, Somerville began work in earnest on her third book, *Physical Geography*, which was first published in April of 1848. After they left England, the family lived a peripatetic existence, moving from place to place as the seasons changed. In addition to Rome, where they spent many winters and some summers, they lived in Florence, Spezia, Naples, Sorrento, Siena, Albano, Turin, and Venice, and visited many other places in Italy. They returned to Great Britain on a few occasions and made occasional trips outside Italy but for the most part lived out their lives there. William Somerville retired from his army post in 1840.

During their residence in Italy, the day-to-day life of the Somervilles and their surviving daughters, neither of whom ever married, remained essentially the same. Mary Somerville spent the morning writing, often in bed, then toured or received visitors in the afternoon and went out in the evening. Though she did not get back to England often, she corresponded regularly with John Herschel and other scientific friends, many of whom visited her in Italy. She also made numerous contacts with Italian scientists as well as with many scientists from other countries who came to Italy for rest or research.

After William Somerville's death in Florence on June 26, 1860, Mary Somerville responded to loss as she had often done before – she began a new book, which was eventually published in 1869 under the title *On Molecular and Microscopic Science*. In her last years, Somerville was somewhat limited physically, but she was intellectually active until the end of her very long life. She continued to receive honors for her scientific work. After she finished *On Molecular and Microscopic Science*, she worked on a number of projects in mathematics. During the last few years of her life, she continued to make the short trip from Naples to Santa Lucia to view the eruptions of Vesuvius. She died quietly in her sleep on November 29, 1872, a little less than a month short of her ninety-second birthday. She is buried in the American cemetery at Naples. Her surviving daughters, Martha and Mary, outlived her by only a few years. Her scientific library was donated to the Ladies College at Hitchin (later Girton College at Cambridge University). In 1879, the second ladies' college at Oxford University was named after Mary Somerville. Originally Somerville Hall, it became Somerville College in 1926. It established a strong reputation within the university and remained a women's college until 1992. (Byrne and Mansfield 1921; Adams 1996)

In the last years of her life, Mary Somerville wrote an autobiography, *The Personal Recollections from Early Life to Old Age of Mary Somerville* (1873), which was completed but not published before her death. The final version was edited by Martha Somerville with assistance from Frances Power Cobbe, a journalist and moral philosopher who had become a friend of the Somervilles in Italy, and John Murray, who had published Somerville's other works. As subsequent chapters will demonstrate, this last book was in some ways more important than her first in determining how she would be remembered.

Toward a Fuller Appreciation of Mary Somerville

If it is true that "we see in Nature what we have been taught to look for" (Nicolson 1959: 5), it is no less true that we tend to see in the history of science what we have learned to expect and to see nothing if the things we expect do not appear. One of the most striking aspects of the foregoing account of Somerville's life is that it bears little resemblance to what we expect in scientific biography – a life organized around recognizable modern disciplines, established institutions, and scientific discoveries of lasting value. This pattern of scientific biography has been dominant until recently, and its underlying assumptions about the criteria for historical significance have been shared by many historians of science. In the cases where women's history has attempted to break out of this pattern, there has often been a tendency to replace one set of stereotypes with another, so that the stereotype of woman as non-scientist is replaced by the stereotype of woman as amateur scientist or woman as popularizer of science.

Although this problem of categories is particularly acute in Somerville's case, it is common throughout women's history and is illustrated in an interesting way in a collective biography called *Pioneer Women*, written by Margaret Tabor and published in 1933. This volume includes biographies of Somerville and three contemporaries whom she knew and admired: the astronomer Caroline Herschel (1750–1847), the actress Sarah Siddons (1755–1831), and the novelist Maria Edgeworth (1767–1849). Tabor's biographical sketches are designed to inspire and encourage young women in intellectual pursuits. A "pioneer woman" is loosely defined as one who is somehow ahead of her time, who accomplishes much and becomes famous, and who also opens new opportunities for other women.

At the beginning of the book, Tabor sets up two categories into which a pioneer woman might fall: the originator and the helpmate. "It may be the lot of a pioneer woman," she tells us, "to win distinction by pursuing some lonely aim and, with little or no help or sympathy from others, to do some splendid work for the world. This is the way of the originator and the leading spirit." (Tabor 1933: 5) This type reflects the mythology that has characterized greatness in science in terms of individuality and independence. Tabor's second category seems equally familiar, especially as a characterization of women's participation in science. "Another may find her mission as a fellow-worker in some great enterprise, as the helpmate, often even the inspirer, of a more powerful initiating mind. . . . In the course of this labour it is not uncommon for the helper to accomplish much and to win fame by her own efforts." (5)

Tabor's first biographical subject, Caroline Herschel, exemplifies this second category very well. Although Herschel became a distinguished astronomer in her own right, she first began to develop scientific expertise in her work as assistant to her brother, William, who was one of the leading astronomers of the nineteenth century. While William pursued a career in music, she supported him in that career and became a soprano herself. When he changed careers and pursued astronomy and telescope making, she moved to astronomy, too. Caroline's primary aim was to help William achieve his goals, even when those goals expanded to include a wife who would take over the domestic duties Caroline had performed. When he died, she felt her own life was at an end.

These feelings notwithstanding, she accomplished a great deal in astronomy both before and after William's death. She is generally recognized as the first woman to have discovered a comet (Alic 1986: 128), and she published extensive astronomical observations. She was awarded the Gold Medal of the Royal Astronomical Society in 1828 and was elected to honorary membership, along with Mary Somerville, in 1835. In the process of helping her brother, she did achieve greatness, but it proved almost impossible for her to identify with her own achievements, and she saw herself as "a well-trained puppy-dog" (125) in relation to her brother. Her attitudes reveal the sometimes pathological dimensions of the ideals of feminine selflessness and humility. She embodied a set of profound contradictions at the same time that she lived out an ideal. The important point here is that, of the four women whose lives are treated in the book, she is the only one whose life seems reasonably if imperfectly captured in the category of the helpmate.

The portraits of the three other women share several common themes. All demonstrate the compatibility of their respective roles in theater, literature, and science with domestic virtues and purity of conduct. All three overcome adversity, endure personal sadness, and go on to achieve success. They all persevere, innovate, and succeed by making contributions in their respective areas and by making their areas of activity more acceptable for and accessible to women. Still, none of them exemplifies the category of the lonely originator. Indeed, all three women are enmeshed in a web of domestic obligation that is often financial as well as emotional; they are surrounded by people who need and depend on them, and they depend on others for support as well. They are hardly alone. And none of them but Herschel fits either of the categories Tabor establishes at the outset.

Tabor's volume illustrates on a small scale what often happens on a much larger scale. The categories she sets up fit well in only a fraction of cases and exemplify influential ideals that are only rarely fully realized. The "originator" category is so deeply rooted in conventional notions of masculinity that it would be virtually impossible for any feminine woman to exemplify it. Because the "helpmate" category is deeply rooted in conventional notions of femininity, it tends to be rather freely applied and to stunt critical attention to the particularities of individual careers and circumstances. Most of the time, the categories fail to capture the salient features of a woman's situation, or they obscure as much as they capture. And they confound these limitations by creating the impression that they have somehow explained a great deal when often all they have done is to force women's scientific activity into acceptable feminine constructs. Mary Somerville is particularly subject to this kind of misinterpretation and is often assigned to categories that reflect some aspect of her career but also distort it or her in some significant way.[9]

One of the principal aims of this book is to establish the distinctiveness of Somerville's career and work and the multifaceted quality of her personality and achievements. Although the categories often distort or mislead, it is still useful to think about Somerville in terms of some of the more commonly used categories. The rather larger number of categories that seem somehow to fit is itself a reflection of the many roles Mary Somerville played. The most common categories reflect the kinds of spaces women tended to occupy, usually because they were seen as being consistent with feminine roles and norms of behavior or because they could be managed in combination with the domestic obligations women had to meet. In many ways, these categories seem to take traditional feminine roles and adapt them to a scientific context.

There are a few categories into which Somerville clearly does not fit. She was neither helpmate nor junior partner in the scientific work of a husband or brother, though she benefited from the mentorship of older scientists and depended greatly on the various forms of support her husband provided. She was also not a marginal or peripheral member of the scientific community, or a writer of science "for the ladies." To conceive of Somerville as an expositor, translator, or popularizer of science puts proper emphasis on the role she played in disseminating new information, ideas, and techniques in science, and emphasizes the role she played as mediator within the scientific community, but it subverts her authority and status as an expert and the formative influence her work exerted. To associate Somerville with the category of the "scientific lady"[10] appropriately recognizes the role that her social acumen and privileged position as a member of the middle class played in making her participation in science possible, but it carries the con-

---

[9] Tabor does not explicitly miscategorize Somerville, mainly because she abandons systematic analysis before she gets to Somerville.

[10] For discussion of the concept of the "scientific lady," see Alic (1986), Phillips (1990), and Schiebinger (1989).

notation of leisure activity, which is inconsistent with Somerville's serious engagement with science.

Somerville stands in an interesting relation to a relatively new category called the "scientific mother." (Gates and Shteir 1997: 9) This category consists of women whose interest in and efforts to promote the spread of scientific knowledge were viewed as part of their responsibilities as good mothers or mother substitutes (such as governesses). Women writing in this tradition often used dialogues, conversations, or allegory in their works and located science in domestic settings. (Gates and Shteir 1996: 9) Jane Marcet, a close friend of Somerville's, is perhaps the best known of the scientific mother figures. Marcet seems to have accepted and relished her role as popularizer and published five widely read books on chemistry, political economy, natural philosophy, and vegetable physiology, which were cast in dialogue form and usually viewed as addressing an audience of women and children. (Alic 1983; Lindee 1996; Myers 1997) Like the scientific mother, Somerville was "an exemplar of female knowledge and authority" (Gates and Shteir 1997: 9), but she never wrote in dialogue, letter, allegorical, or conversational form, and she never conceived of herself or was seriously perceived as writing for women and children. She addressed expert readers as equals and collaborators and did not assume a maternal relationship to her readers. All this said, it seems likely that her resemblance to the figure of the scientific mother helped Somerville gain acceptance and manage her identity as a female authority figure.

Somerville could also be mistaken as a purveyor of what Barbara T. Gates and Ann B. Shteir term "the narrative of natural theology," in which "each discovery of natural science was new proof of the wisdom and power of a divine creator." (11) This narrative tradition associated scientific observation with moral education, emphasized anecdotes, and minimized theory. It was strongly associated with women, especially in the early nineteenth century, but men certainly participated in it as well. As Somerville's use of the scientific sublime and other theistic themes demonstrates, she accepted many of the basic propositions of natural theology and derived additional pleasure from perceiving the workings of the creator in the natural world. But proving the existence of the creator was not her purpose in writing about science and not her vision of the purpose of science. Although the notion of science as a pathway to God was an integral element in the interpretive framework within which she located science, natural theology occupied only a very small amount of space in her writings.[11]

---

[11] There are a number of categories that come close to accurately representing Somerville's career in science. One of these is "expert popularizer," which captures the

A number of other categories not typically associated with women do accurately reflect some of the important roles Somerville played. One of these is "historian of science." Though she rarely develops any kind of extended historical narrative, she does use the history of science to provide conceptual background and a sense of motivation, aims, and methods for different areas of scientific inquiry. She also considered which discoveries would be judged of most lasting value. But this historical work was a significant minor thread in her career rather than a major theme.

One of the most useful categories for understanding Somerville's career and influence is "organic intellectual." This category was originated by Antonio Gramsci and was used by Deirdre David (1987) in her analysis of Victorian intellectual women. Gramsci argues that every rising social group "creates together with itself, organically, one or more strata of intellectuals which give it homogeneity and an awareness of its own function." (1971: 5) These organic intellectuals require a range of expertise and ability to organize and operate in a number of spheres. They are not, in Gramsci's view, merely eloquent orators, but also active participants in practical life. Organic intellectuals are distinguished from the traditional intellectuals who evolved along with the existing dominant class and have become entrenched. Though she had strong links with the traditional intellectuals of her day, Mary Somerville clearly fits the pattern of the organic intellectual, in the sense that she developed alongside and was encouraged by the members of the middle class who sought to change society and secure their positions through the promotion of science and other reforms.

Somerville played an instrumental role in developing and disseminating a vision of science that helped promote cultural unity and that located science within the context of a liberal social agenda. In keeping with the pattern that David (1987) posits for intellectual women, Somerville both accepted and resisted middle-class ideals, especially as those related to women.[12]

---

fact that her expertise was acquired through a small amount of her own experimental and observational work and a large amount of first-hand contact with the leading scientists of the day. It also reflects the fact that she addressed scientists as equals and was treated as one by them. The term "popularizer" is so vague and so loaded with negative connotations that I prefer not to use it at all. For an interesting discussion of the concept of the expert popularizer and for an interesting comparison with Somerville, see Lightman (1997) and his discussion of the work of Agnes Clerke. There are many parallels between the two women, perhaps the most significant of which is that Clerke also addresses experts as an expert.

[12] Because most of her intellectual output was concerned with the structure and meaning of scientific ideas, she might more appropriately be called a "scientific intellectual."

David sees intellectuals in the nineteenth century as taking on the job of counseling "the middle class in matters of economic, social, and religious change." (11) In her writing Somerville undertook this function, which was consistent with the notion of woman as moral guide, though always as a frame for a central discourse about the natural world as it was revealed through science. Because of her scientific expertise, Somerville was able to clear up "uncertainty about scientific change" (11) and its meaning. As in the case of her relation to the scientific mother, Somerville benefited from being cast in a culturally respectable role that allowed her to combine intellectual pursuits with a feminine identity. Somerville's greatest resistance to the prevailing norms for women was that she did not see the object of her intellectual development as part of her role as a wife. She pursued science to develop her own ability, pursue her own curiosity, and demonstrate that women were capable of assuming a higher place in the intellectual world than that usually assigned to them. She enjoyed the companionship and support of her husband in the effort, but her career was not coextensive with her marriage.[13]

Although some of these less commonly used labels seem to work reasonably well, I have concluded that it is most appropriate simply to call Mary Somerville a "scientist" and to define the term broadly. It is telling that the first published use of the term "scientist" occurred in a discussion of Somerville's work, Whewell's 1834 review of *Connexion*. He had proposed the term at the 1833 meeting of the British Association for the Advancement of Science. In the review, he establishes it as a term analogous to "artist" and "economist" and defines it broadly as a "name by which we can designate the students of the knowledge of the material world collectively.... a general term by which these gentlemen [of the BAAS] could describe themselves with reference to their pursuits." (59) The term did not catch on immediately and did not really take root until the process of professionalization was complete and science had come to be viewed as "the main characteristic of the age" (Merz 1965, 1: 89) at the end of the nineteenth century.[14]

The shifting meanings of the term notwithstanding, there are, it seems to me, several compelling reasons for calling Mary Somerville

---

[13] Though the Somervilles definitely had a strong partnership and what Peterson (1989) calls a "companionate marriage," they differ from the couples Peterson discusses because the scientific career in question was primarily hers.

[14] The term did not take on its current narrow definition until the process of professionalization was complete at the end of the nineteenth century. For an example of the ways in which narrow definitions of the term "scientist" and the use of dualistic thinking can distort Somerville's accomplishments even when the scholars in question seek to raise her profile, see Toth and Toth (1978).

simply "a scientist." These include: (1) her high level of expertise, (2) the extent of her participation in the scientific community of her day, (3) her concrete contributions to the advancement of science, (4) the recognition she achieved both from her peers in science and from the British government (in the form of a government pension), (5) the fact that her work was judged by the same standards as that of her male colleagues, (6) the fact that science was both her vocation and a significant source of her family's livelihood, and (7) her qualities of mind. This definition of the term "scientist" emphasizes expertise and innovation rather than the terms "originality" and "discovery," which are both strongly associated with masculine notions of creativity.[15] It also emphasizes the social dimensions of science more than the individualistic ones.[16] In all of these respects, such a broad definition more fully accounts for the majority of people who are considered to be "scientists" at any point in history. My definition also relies heavily on conceiving of science as an enhanced form of perception and expression as well as a form of knowledge or social activity. The term "scientist" seems particularly appropriate when we look at Somerville through the eyes of her contemporaries, who judged her by the standards of her own time rather than by the more limited criteria of ours.[17]

[15] For a discussion of the gender implications of concepts of originality, see Carroll (1990). Although few examples are discussed in this book, the word "genius" was often used in connection with Mary Somerville.

[16] For an interesting discussion of the role of individualism in British science, see Merz (1965: 287). Merz also has a very interesting discussion of naturalism in relation to individualism that helps put Somerville in a broader context.

[17] The definitiveness, precision, and stability we associate with scientific knowledge often obscure the nebulosity of the complex of activities, values, and meanings we term "science." One term that captures the complexity and layers of activity involved in science is "metascience." See Yeo (1993) for a discussion of the concept of the metascientist as he develops it in his study of William Whewell. Metascientific activity takes a wide range of forms and pursues a number of aims, which might be broadly categorized as (1) promoting science, establishing its value, and justifying support of it; (2) obtaining institutional support for science by creating new institutions or securing a place for science within existing institutions; (3) delineating science, defining and distinguishing it from other kinds of knowledge and activity; (4) interpreting science for a wider audience, relating it to other cultural values and activities, and establishing a consensus about the role science should play in society; and (5) evaluating or assessing the actions or products of individual scientists or the scientific community, reflecting critically on science, on its history and method, and on the conditions conducive to its advancement. Somerville was very much involved in many of these activities and might herself be called a "metascientist." I have not used the term extensively in my discussion of Somerville because it has not been widely accepted.

It is not much easier to find a term to describe the larger group or groups of which Somerville was a part than it is to find a label for her. She was a close associate of many of the "Gentlemen of Science" whose activities are the subject of Morrell and

Omniscience as an Intellectual Ideal

Once we get past the obstacles created by using inadequate categories and overly narrow definitions of science, we face those created by a failure to recognize the differences between Somerville's time and our own. Most of these differences are well delineated in recent historical scholarship, but they are not always acknowledged in the evaluation of individual careers. Somerville's eminence and achievements become comprehendible only when we understand her world in its own terms. One of the most important differences between her time and ours arises from the fact that at that time science was not yet professionalized and highly specialized; omniscience prevailed as an intellectual ideal.

Although the dualistic framework that divided the world into separate spheres of masculine and feminine, public and private, was well established, and, indeed, pervasive in Somerville's time, and science had already taken on a masculine identity, the dualism of professional scientist versus amateur scientist was not yet established and had not been mapped onto gendered distinctions. The professional structures of degree programs, professorships in science, large laboratory-based research, and peer-reviewed journals, that would, later in the century, define a career in science, were not yet fully elaborated for either men or women. Science was not conceived as either "a profession" or "a consuming interest and art," by Somerville and her contemporaries, most of whom pursued science in addition to a number of other intellectual activities. There were, of course, distinctions made in levels of scientific expertise, but there was not yet a duality posited between popular and professional writing.

Another aspect of focusing on specialization is a tendency to equate the serious pursuit of science with a high level of specialization. This tendency is particularly important to recognize because the breadth of subject matter Somerville wrote on might be taken to suggest

Thackray's study of the origins of the BAAS (1981). Cannon (1978) includes her as member of "the Cambridge network." Somerville's sex makes it seem incongruous to call her a "Gentleman of Science," and the term "Cambridge network" is not widely used or understood. I have decided to designate the various individuals and groups with whom she was associated as "cultivators of science" in the sense that Whewell used that term. This term allows for many kinds of involvement with science but reflects two important characteristics: (1) an association with encouraging, fostering, furthering, nurturing, and promoting the cause of science, rather than just observing it, and (2) its consistency with the perception of that time regarding the need to integrate science more fully into culture.

dilettantism. Today, we find it hard to believe that any one individual could master such a broad range of knowledge, or that it could be done in a way that would win the respect of what is ordinarily called "the scientific community." This reasoning reflects a trap created by two mistaken assumptions: (1) that specialization has always been the dominant ideal, especially in science, and (2) that the barriers between "the two cultures" have always been defined and conceived in the same way.[18]

Careful examination of history reveals that early Victorian intellectuals thrived on controversy and debate but took a unitary approach to intellectual life. They saw a culture as an organic whole and were ambivalent about specialization because of the division of intellectual life that it encouraged. (Schweber 1981) Somerville's intellectual circle was not made up entirely of scientists. Although her peers tended toward mathematicized physical science, the group included a wide range of scholars who believed that the goal of education was knowing "something of everything and everything of something." (De Morgan quoted by Schweber 1981: 3) The ideal of the group was polymathy or omniscience, reflected in a craving to "master all that was known and could be known." (3) They saw this ideal as difficult but attainable and expected that all members of their group would strive for it.

While it may seem contradictory from our point of view, they placed a high value on generalist education at the same time that they recognized the value of professionalization and specialization. They believed that producing new knowledge was essential but were not willing to give it primacy over the acquisition and dissemination of knowledge. This unitary ideal was gradually abandoned as science became increasingly professionalized, as knowledge expanded rapidly, and as disciplines multiplied. But in Somerville's day the ideal of omniscience prevailed, and intellectuals whose primary interest was in science were still very much integrated into the broader intellectual community.

Somerville's friend John Herschel declined the presidency of the British Association for the Advancement of Science when he was first offered it because he feared the BAAS would encourage the compartmentalization of science; similarly, he declined a Lucasian Professorship because he thought it would make it difficult for him to pursue the full range of scientific and intellectual pursuits in which he

---

[18] For a discussion of the history of the configuration of the disciplines see Collini (1993).

was interested. These beliefs about knowledge and specialization were rooted in particular views of the natural and social worlds and bore an interesting relation to the problems British society faced at the time.

The industrial and political revolutions begun in the eighteenth century had contributed to social fragmentation; a new vision that would "animate the hopes of the various social classes" was needed. (Schweber 1981: 7) For Somerville and other organic intellectuals of her day, the answer was a progressive view of culture in which science played an essential role. But science could serve as the focus of a unifying vision only if it clearly related to other aspects of culture and supported moral, social, religious, political, and material goals. Creating and promoting such a unifying vision had both an intellectual and a rhetorical component. It required both broad and deep knowledge and persuasive skills. Somerville possessed these to a high degree, and her ability to synthesize and bring disparate elements together was highly valued. The task of creating a unifying vision was facilitated by the philosophical commitments that she and many of her contemporaries shared.

The philosophical basis for the group's willingness to recognize "several kinds of excellence" and their tendency to shun "departmentalization for its own sake" (Cannon 1978: 30, 58) was a unified view of the world that was primarily religious in origin. As Susan Faye Cannon paraphrases this view: "After all, no one of the actual phenomena of God's world is described exhaustively by one scientific specialty; each one could be seen fully only if seen in its relation to all the others." (58)

There were also practical motivations behind the desire to amass factual material. Somerville and her contemporaries believed science advanced by achieving increasingly higher levels of generalization. These higher levels of generalization usually took the form of new laws with greater explanatory power. The early Victorian scientists were motivated not by mindless fascination with detail but rather by "a desire for more and more widely applicable laws to interconnect the diverse phenomena, not merely laws of chemistry, or magnetism, or geology, but, for example, laws of electro-chemistry, electro-magnetism, of physical geology [with] . . . an interest in the detailed complexity of the actual." (Cannon 1978: 58–59) Increasingly higher levels of generalization could be achieved reliably only through the accumulation of increasingly large amounts of detailed information and the search for meaningful patterns and analogies. The task of generating or observing such enormous amounts of detail might be too much for an individual, but an individual might aspire to accumulate and assimilate massive amounts of detail from which important patterns and useful analogies emerged. The result of this set of perspectives was that each person "saw his work not as an isolated bit of a specialty, but as a part of an intellectual totality." (63)

This is not to say that the seeds of the "two-cultures" distinction did not yet exist, because they most certainly did. As the foregoing discussion demonstrates, the insistence on a unitary vision was at least in part a response to a perceived fragmentation of intellectual life. In his 1834 review of *Connexion*, Whewell spoke with alarm of the increasing tendency of the sciences toward "separation and dismemberment." He harked back to the day when all learned people "embraced in their wide grasp all the branches of the tree of knowledge" and lamented "the inconveniences of [the] division of the soil of science," which he referred to as "an evil." (59–60) Regretfully, he concluded, "these days are past; the students of books and things are estranged from each other in habit and feeling. . . . We adopt the maxim 'one science only can one genius fit.' " (59) One of the reasons that Whewell so much admired Somerville's work was that it was dedicated to the "noble object" of demonstrating the unity of the sciences. This objective had a decidedly aesthetic and spiritual component but also reflected a belief about how science advanced and progressed. The person who possessed a comprehensive view was in a better position than persons of narrow views to perceive "connexion, dependence, and promise" (54) and to help advance science.

As Stefan Collini (1993) has pointed out, one of the underlying themes of the "two-cultures" debate is a concern about the possibility of maintaining the capacity for "debate or mutually intelligible exchange of views upon which the conduct of society's affairs depends" in the face of the specialization that seems prerequisite for intellectual progress. (lvii) Somerville's work helped establish and maintain the conceptual and rhetorical space in which meaningful exchange could occur, both between scientists in various specialties and between emerging communities of experts and those whose scientific expertise lay in other areas.

The Scientific Sublime as Blend of Aesthetic, Religious, and Scientific Elements

Other roadblocks to accurately characterizing Somerville arise from a tendency to underestimate the importance of the rhetorical dimensions of science and to see scientific discourse as isolated from other forms of discourse. Contemporary research scientists tend to view the process of "arranging one's findings in intelligible form . . . as something of a chore" and as disconnected from the process of discovery, rather than as creative activity. (Collini 1993: lix) Recent years have seen a growing recognition of the links between science and literature, but the overall tendency is still to view scientific discourse as isolated from creative activity and from other kinds of cultural discourse. In Somerville's day, the unitary character of intellectual life was reflected in the blending of various forms of discourse in the writing of Somerville, Whewell, Herschel, and many other influential figures in the scientific community of the time. This blending was common even in writing whose primary purpose was scientific.

To understand the strength and impact of Somerville's work, we must also get past the ideas that science stands in an adversarial relationship to poetry and to religion and that science is somehow diminished or tainted by the introduction of poetic or religious elements. Like the other roadblocks mentioned here, this one is widely recognized by specialized scholars but still stands as a formidable obstacle in the minds of many observers.

From a historical point of view, it is clear that the early Victorians held quite diverse opinions about exactly what the relationships among science, religion, and poetry were or should be. But there was also a common belief that science was very much intertwined with both poetry and religion. And the three enterprises were seen as being at least potentially mutually supportive of each other. Scientific intellectuals in Somerville's day promoted science by attaching religious, moral, and aesthetic as well as practical and intellectual value to it.

Science did not begin to be perceived as "tainted" by religious beliefs until the later Victorian period. (Lightman 1988; Zaniello 1988) During the period when Somerville made her reputation, scientific writing that did *not* have some kind of reference to the creator would have been highly suspect and viewed as unsafe science. (Topham 1992) The prevailing belief was that science could not be taught well without reference to the sublime truths of natural theology and that the scientific study of nature revealed God. Most of early nineteenth-century gentlemen of science treated their endeavor as what Jack Morrell and Arnold Thackray (1981) have described as "a Godly but autonomous intellectual pursuit." (230) Somerville treated science in a similar way.

Moreover, science was in the process of justifying itself in an intellectual and institutional environment that was dominated by classical literature and by individuals who saw knowledge of literature and the possession of a refined literary style in writing as the hallmarks of an educated person, regardless of that person's profession or intellectual specialty. Literary references, well-turned phrases, and a writing style that tended on occasion to call attention to itself were more or less required for any writer who wished to be taken seriously. Somerville, like many other scientific writers, tended to use a restrained literary style that drew on rich literary traditions to achieve its aesthetic effect.

One of the greatest benefits of getting beyond these last two roadblocks is that it becomes possible to use literary history to shed light on the techniques Somerville used and the sources of power on which they relied. The scientific sublime that Somerville was so skillful at evoking is but one variant in one of the most important and complex aesthetic categories in the Western tradition, and is not easily reduced to a set of essential traits. Andrew Ashfield and Peter de Bolla (1996) have suggested that the sublime has been the most common rubric for answering the question: "What is it that moves me?" (21) In a somewhat narrower vein, Thomas Weiskel (1986) has argued that "The essential claim of the sublime is that man can, in feeling and in speech, transcend the human" and link personally with the great or the superhuman. (1) In the scientific sublime, the reader links with the great in the form of an encounter with the attributes of God revealed in nature by science.

The scientific sublime, then, is a response to nature, but it is very much a mediated response, and the greatness the reader experiences is in part the greatness of intellect and grasp represented by scientifically informed perception, which dramatically expands the frontier of the unknown. The scientific sublime thus differs from the natural sublime in its emphasis on science as a tool of perception, and it also differs from the Romantic sublime. Another way to put this is to say that the scientific sublime is a response to both the nature perceived through science and the technique of perception, that is, science itself.

Unlike treatments of the traditional sublime, Somerville's writing makes it clear that the most significant enlarged power offered by mathematical science, especially as it was exemplified in physical astronomy, was the predictive power that allowed human beings to make reliable predictions based on relatively small amounts of direct observation. This was not a power of nature, but a power of intellect embodied or made manifest in science. Through their creation of the natural sublime, the poets had already provided the basis for the powerful rhetoric that Somerville used to frame the presentation of the substance of science and for constructing appealing views of the world based on science.

The scientific sublime, thus, was a powerful blend of aesthetic, spiritual, and scientific elements. Somerville did not create the literary tradition that combined these elements. In fact, as the prologue to this book suggests, its power arose from its familiarity and pervasiveness. Somerville was certainly not the only prose writer on science to adapt the themes and approaches of scientific poetry, but she was certainly one of the most skillful and successful in presenting science as a form of elevated meditation.

Understanding Somerville's work in relation to the history of the scientific sublime and of scientific poetry reveals the coherence and continuity in Somerville's works when they are considered as a whole. William Powell Jones argued that the sublime disappeared from scientific poetry at the end of the eighteenth century because the poetry collapsed under what might be called "the didactic burden." When poets tried to convey detailed scientific information, the poetry became what Jones calls "tedious technical versifying." (Jones 1966: 205) If it is true, as he argues, that "too much knowledge is bad for poetry," then Somerville's work demonstrates that a little poetry can also be very good for scientific prose.

Jones also suggests that the idea of science as a pathway to God – a way of perceiving the divine wisdom in nature – had been rejected by philosophers by the end of the eighteenth century. He implies that the themes of the scientific poets had somehow also been discredited or fallen out of favor, an implication that is not borne out by examination of Somerville's scientific writing or the writing, both poetical and otherwise, that appeared among persons writing on science in the nineteenth century. For example, the themes of the scientific poets are clearly at work in Sotheby's *Lines Suggested by the Third Meeting of the British Association for the Advancement of Science* (1834), where he refers to scientists as leading people both intellectually and morally, to "Sublimer views," refined moral sense, and "from earthly wisdom to divine." While I would not argue that this is sublime poetry, I would contend that it conveys the vision that animated much of early to mid-nineteenth-century science. Somerville's scientific writing demonstrates that the aesthetic and intellectual traditions associated with the scientific sublime had not died; they had simply found a new medium for expression, a medium more capable of shouldering the didactic burden.

Subsequent chapters will examine Somerville's individual works in detail. But it may be useful to follow the example she set in her own writing and to consider her works as interrelated parts of a whole before examining them in their individuality and particularity. This is especially appropriate because the works overlap and connect in many ways and because the order in which she took up subjects and directed her attention to the natural world closely parallels the sequences followed by the scientific poets.

The "Preliminary Dissertation" to *Mechanism of the Heavens* (1831) became an independent publication in 1832 and expanded further to become *On the Connexion of the Physical Sciences* (1834) and its subsequent editions of 1835, 1836, 1837, 1840, 1842, 1846, 1849, and 1858.[19] Having begun with the heavens and physical astronomy, Somerville turned in *Physical Geography* (1848) to the earth – its physical features and inhabitants, vegetable and animal. She continually updated her readers on the latest discoveries on these topics through successive editions published in 1849, 1851, 1858, 1862, and 1870.[20] In her last work, *On Molecular and Microscopic Science* (1869), she turned to the molecular and microscopic worlds. For this last work, she adopted Augustine's motto, "Deus magnus in magnis, maxima in minimis" – God is great in great things, greatest in the least. She had begun with the great things, concluded with the least, and covered a vast terrain of knowledge in between.

Thus, as she moved from *Mechanism of the Heavens* and *On the Connexion of the Physical Sciences*, through *Physical Geography* and *On Molecular and Microscopic Science* she followed the same succession of subjects that both Nicolson and Jones discerned in their studies of the scientific poets. Nicolson sees a pattern of development that moves from the "cosmic voyage" to the "terrestrial excursion": "The vastness of God, earlier read over into the vastness of cosmic Space, is now reflected in the vastness of a world filled with all possible variety, a world as engrossing to human imagination as the cosmic heavens had been." (Nicolson 1959: 293–294) Jones sees a similar pattern, in which the poets moved from the world revealed by the telescope to the world revealed by the microscope, from physical astronomy to natural history to landscape description of flowers, animals, and birds. As Jones evaluates these changes, the moral purpose and "religious themes continued, throughout the century, but the scientific examples used to illustrate God's presence in nature changed from celestial systems to

---

[19] A tenth edition edited by Arabella B. Buckley was published in 1877, after Somerville's death.
[20] A seventh edition of *Physical Geography* was published in 1877.

English flowers and birds" and "attention shifted to ... the gentler world of nature that could be observed close at hand." (192, 232) One difference between Somerville and the scientific poets is that she was able to evoke the scientific sublime throughout the full range of subjects.

One way to conceive the development of Somerville's writing over time is to see her working her way through the cosmos from the glories of the celestial sphere to the equally glorious realm of the microscopic sphere. Her movement over time and through space suggests that one key aspect of the illuminated female mind – the capacity to add to "the great merit of being profound ... the great excellence of being clear" – is the ability to move easily between the cosmic and the everyday and to use each of the realms to enrich our appreciation of the other.

Conclusion

In the end, there is no simple answer to the question: Who was Mary Somerville? As the preceding discussion demonstrates, she played not just one role but many, had expertise in not just one science but many, wrote not only for one audience but for several. She was both mathematician and poet in soul, model of ideal womanhood and respected scientific intellectual. This blending of presumed opposites might seem surprising, but at a second glance also appears to be characteristic of innovative individuals and works.

In his *Philosophical Inquiry into the Origins of Our Ideas of the Sublime and the Beautiful* (1759), Edmund Burke set up an apparent duality that opposed the sublime to the beautiful. He used terms that are not explicitly gendered but easily construed as gendered. He defines the sublime as vast, rugged and negligent, solid and massive. The beautiful, by contrast, is small, smooth and polished, light and delicate. Interpreters often treat these two categories as entirely separate qualities. Although Burke granted that there was "an eternal distinction between them," he also argued that "we must expect to find the qualities of things the most remote imaginable from each other united in the same object." (Burke in Ashfield and de Bolla 1996: 140) He was, in this context, thinking of objects in nature and art, but his analysis can be applied to individuals as well. The ability to unite apparent oppositions formed a great deal of Somerville's strength as a scientist and her appeal as a public figure.

This chapter has identified obstacles and useful historical perspectives in an effort to make it easier to appreciate and define Somerville's accomplishments. It has also presented a rudimentary account of her life story. In the process, it should have created a framework for more fully appreciating Somerville and her work. The next chapter moves beyond the mostly uninterpreted narrative presented here to a more analytical account that attempts to assess the factors that were most crucial to Somerville's development and success.

2

Creating a Room of Her Own in the World of Science

How Mary Fairfax Became the Famous Mrs. Somerville

Early associations never entirely leave us, however much our position in life may alter.
— Mary Somerville, *Physical Geography*, 1848

The more one learns about Mary Somerville's childhood and early life, the more surprising her accomplishments appear. Her father had been sent to sea at the age of ten and had little formal education. Her mother read only the Bible, the newspaper, and sermons. Her parents' educational goals for their daughter were for her "to write well and keep accounts, which was all that a woman was expected to know" (PR 25) and to learn what might be called the domestic fine arts: needlework, pastry and jam making, gardening, and piano. As a child, she had great difficulty remembering names and dates and found the catechism incomprehensible. Though she could read *Pilgrim's Progress* at the age

45

of eight or nine, she did not master the basics of writing and arithmetic until she was thirteen years old. She never became adept at spelling or doing simple sums.

Her family and friends generally disapproved of her penchant for reading, and she was prevented from studying Euclid at night because she was depleting the family candle supply and her family feared she would follow in the footsteps of an acquaintance who had gone "raving mad about the longitude!" (PR 54) In girlhood, her education was inhibited by old-school prejudices, lack of money, and the demands of acquiring and practicing the domestic and artistic skills required of a woman of the middle class; in early adulthood, she was limited by isolation and an unsupportive spouse. She had only one year of formal academic instruction, and that was thoroughly miserable and almost entirely unproductive. In short, her experience was very much like that of the majority of middle-class women of the late eighteenth century and early Victorian period. Yet she became a highly accomplished mathematician and one of the leading scientific authorities of her day. All of which raises perhaps the most interesting question about Mary Somerville: How did she do it? How did Mary Fairfax become the famous Mrs. Somerville?

Her strategy reflects the overall pattern of resistance and acquiescence identified by Deirdre David in *Intellectual Women and Victorian Patriarchy* (1987). That is, she both accepted and selectively resisted the norms imposed by male-dominated culture, which both undermined and encouraged her, and which granted her authority as a scientific intellectual but only on the condition that she accept significant limitations. The account that follows identifies the crucial choices Somerville made to advance in her career as an intellectual; it specifies the respects in which she resisted and the ways in which she acquiesced to the larger system of which she was a part. It attempts to strike a balance between giving due credit to Somerville's talent, effort, and perseverance, and demonstrating the role played by both fortuitous events and the support network in which she operated, a network that put her in a position not only to do significant scientific work but also to be recognized for that work. In particular, it highlights the complex of factors that contributed to her success and the ways in which she took advantage of the spaces – the opportunities – that were available, how she defined, expanded, and used them. Overall, her capacity to create a room of her own in the world of science resulted from a combination of ability and good fortune with determination, flexibility, and good instincts regarding the social dimensions of intellectual life.

The Gender Landscape of Science in the Early Nineteenth Century

In the era when Somerville was pursuing her education and making her reputation (roughly 1790 to 1835) there was fairly widespread support, at least in principle, for the education of middle-class women, but the education available tended to be limited, somewhat haphazard, and based on assumptions of feminine inferiority in intellectual matters. It was also shaped by the presumption that the sole aim of women's education was to prepare them for their place in the domestic sphere and their roles as wives, mothers, and guardians of morality. (David 1987; Corbett 1992) In the seventeenth and eighteenth centuries, science had come to be seen as an acceptable, indeed fashionable, form of amusement among the upper classes, but the serious pursuit of science was generally believed to be beyond the capability of women or detrimental to their femininity, or both. This created a situation in which science was pervasive but at the same time beyond the reach of girls and women at anything except a superficial level. It was encouraged, but only in a very circumscribed way.

On the other hand, there were also respects in which the serious pursuit of science was more available to women than it would be later in the nineteenth century and well into the twentieth. Somerville established herself during the transition from natural philosophy to natural science that occurred in the late eighteenth and early nineteenth century. Science was being established as a distinctive form of knowledge and its practitioners were attempting to establish a cultural base for it. Women's historians have recognized a pattern in which women managed to "gain a foothold [in] new fields of science, which lacked fully elaborated career structures." (Rose 1994: 101) In Somerville's time, especially early in her career, this characterization could be applied to science generally, not just to particular fields. Science had not yet professionalized in the sense that it was still primarily an amateur activity in Great Britain and there were very few full-time, paid science positions in institutions. Though experimental and theoretical work required considerable expertise, most productive scientific practitioners worked in domestic settings, depended on private patronage (instead of government support), and relied on family members to assist them in their work. Science was more private than institutional, and it could be said that letters of introduction were at least as important as university degrees in such an environment.

Narrow specialization was not yet the dominant model for the acquisition of credentials or the conduct of scientific careers, and many practitioners developed their expertise through informal apprenticeships or independent study. This was true even for those who had attended university, largely because many of the disciplines we now think of as the physical sciences had not yet assumed a secure place in university curricula. Somerville conducted her published experimental work on the effect of the solar spectrum on plants in her home with very simple apparatus. In a letter of November 21, 1845, John Herschel spoke approvingly of the simplicity and elegance of her experiments, which were undertaken with minimal equipment in a domestic context. (PR 278–279) The main external resource she needed was extended sunny weather, which was available in abundance in Italy and during at least one fortunate summer in England. There were, of course, limits to the kinds of discoveries Somerville could make in such a context, but there were fewer institutional barriers to the resources required to practice science.

The net result of these differences was that "the experience of involvement in science was not as radically demarcated between the sexes as it would become later in the century." (Abir-Am and Outram 1987: 2–3) Like other women of her generation, Somerville had to confront entrenched prejudices often expressed as a "dread of learned women," as well as formidable social conventions regarding the abilities and qualities that she was required to cultivate to maintain attractiveness and acceptability as a woman. (Alic 1986, David 1987) But while she faced social pressure and psychological barriers to her pursuit of science, institutional obstacles were of less consequence in Somerville's time than they would become later. Most of Somerville's battles were fought and her network of associations was formed in a domestic context. Although her sex meant that it was impossible for her to attend a university or become a fellow of the Royal Society, she had full access to the social circle in which the leaders of those organizations moved. She could be taken seriously as a scientific intellectual without being eligible to hold either a faculty or student position. The psychological barriers that Somerville faced were less visible than institutional barriers but not less powerful since they were most often internalized as well as imposed. The story of Caroline Herschel shows how crippling these internalized barriers could be, and Somerville's story shows how hard these barriers could be to overcome. They were in a sense the problems that faced all women involved in serious intellectual pursuits, and, though they were certainly manifest in science, science was not considered a special case to the extent it would be later in the century.

Thus, when Somerville was establishing herself, it was difficult but still possible to practice science seriously and successfully in the domestic sphere, largely because science had not yet entirely "gone public," which primarily meant becoming professional and institutional. When it did, beginning around the middle of the century, most women who were active in science would be left behind in the domestic sphere, relegated to amateur status. (Gates and Shteir 1997; Rose 1994) By that time, Somerville's reputation was well established, and it was so strong that she was able to maintain it for the rest of her long life. Still, she was no longer part of the inner circle of science and might not have been able to remain so even if she had stayed in England, so much had the gender and institutional landscape changed.

In Her Father's House

In the introductory essay to *Personal Recollections*, Martha Somerville offers this portrait of Mary Somerville's early days:

> As a lonely child, she wandered by the seashore, and on the links of Burntisland, collecting shells and flowers; [and] spent the clear, cold nights at her window, watching the starlit heavens, whose mysteries she was destined one day to penetrate in all their profound and sublime laws, making clear to others that knowledge which she herself had acquired, at the cost of so hard a struggle. (PR 2)

Whether Somerville was *destined* to penetrate the mysteries of the heavens is open for debate; her writing clearly reveals that she did retain throughout life the perspective of the child wandering the seashore and contemplating the heavens, and she began by being an astute and avid but naïve observer of the natural world. She never forgot what it meant to find phenomena mysterious or impenetrable.

Martha Somerville's portrait of her mother's early life was often repeated in accounts of Somerville's life, in part because it resonates with the sense one gets from Somerville's scientific writing, and perhaps also because it echoes Isaac Newton's well-known statement:

> I do not know what I may appear to the world; but to myself I seem to have been only like a boy, playing on the seashore, and diverting myself now and then finding a smoother pebble or a prettier shell than ordinary, while the great ocean of truth lay all undiscovered before me. (Newton in Brewster 1860, 2: 331)

Among their many possible meanings, these statements reflect a humility in the face of the immensity of creation that was an article of faith with both Newton and Somerville. They also paint a picture of an intellect that grasps both complexity and immensity but never loses interest in the mundane aspects of nature. As the child grows to adulthood, the world still seems immense because the individual's breadth of conception is so great. This kind of intellect retains a sense of childlike wonder, at least in part because it is able to grasp reality on a cosmic scale.

Much of Mary Somerville's success as a thinker and a writer, especially her capacity for synthesis, was rooted in the ability to hold a great deal of information in memory, to contemplate it with what might be called "the mind's eye," and to present her vision to the reader. Like her overall character and career, Somerville's imagination resists typecasting, especially in the terms of Romanticism, which tends to oppose the creative aspect of imagination to reason and the kind of analytical thinking associated with science. (Hill 1977: 12–13)[1] Somerville's imagination combined the power to represent what is absent with a capacity to combine and associate ideas, to synthesize disparate elements, and to permit "the mind to see beneath the transitory surface of the material world." (13) Her imagination might be characterized as synthetic and revelatory, and consistent with the general role that science played in expanding human intellectual horizons. The kinds of oppositions between reason and imagination that discussions of Romanticism often suppose do not manifest themselves in Somerville's imaginative treatment of science.

In her earliest years (roughly 1782–1791), she stocked and developed her imagination and developed both an affinity for and an intimate acquaintance with the world of nature, both terrestrial and celestial. Throughout her life, the early associations acquired during her childhood at the quiet seaside village of Burntisland were the strongest. Though she mastered the intricacies of analysis and grasped the sublime truths revealed through physical astronomy, the environment and the manner of life that she, her mother, and her brothers pursued at Burntisland had the profoundest effect in developing the qualities she retained throughout her life and conveyed in her writings. Solitude and intense interaction with nature were the hallmarks of her early years.

The detail in which she describes her childhood home reveals both the quality of her powers of observation and the strength of her memory. She tells us in *Personal Recollections* that the Links, a grassy area on the east of town of Burntisland,

> were bounded on each side by low hills covered with gorse and heather, and on the east by a beautiful bay with a sandy beach, which, beginning

---

[1] It is also worth noting that Somerville's work does not conform to Mellor's (1993) notion of the domestic sublime, which she associates with feminine Romanticism.

at a low rocky point, formed a bow and then stretched for several miles to the town of Kinghorn, the distant part skirting a range of high precipitous crags. (PR 10)

The house they inhabited had a southern exposure and extensive gardens of fruits, vegetables, and flowers, which were "bounded on the south by an ivy-covered wall and hid by a row of old elm trees, from whence a steep mossy bank descended." (PR 11) Burntisland seems to have provided an ideal setting for the development of the strong aesthetic sense and the keen eye for observing nature that so strongly pervade Somerville's scientific writings.

Because her brother Samuel spent most of his time in Edinburgh with their grandfather, Somerville spent most of her early childhood alone in the garden of the family house. Although her statement that "I never cared for dolls, and had no one to play with me" (PR 18), might be taken as an early rejection of domesticity, it is probably more accurately interpreted as a reflection of her fascination with the natural world and the appeal of the freedom she came to associate with it. Somerville also acquired a love of birds at Burntisland. Many were allowed to feed near the house and on the Somerville property, as well as to build nests above the windows of the house. This interest stayed with her throughout her life. Of the birds she says, "I knew most of them, their flight and their habits. The swallows were never prevented from building above our windows, and, when about to migrate, they used to assemble in hundreds on the roof of our house, and prepared for their journey by short flights." (PR 18)

After her year of confinement at boarding school (c. 1791), she returned to Burntisland with a renewed appreciation for its virtues. "I was," she says, "like a wild animal escaped out of a cage." She wandered farther than she had before, and studied what she found with greater intensity.

> When the tide was out I spent hours on the sands, looking at the starfish and sea-urchins, or watching the children digging for sand-eels, cockles, and the spouting razor-fish. I made a collection of shells, such as were cast ashore, some so small that they appeared like white specks in patches of black sand. There was a small pier on the sands for shipping limestone brought from the coal mines inland. I was astonished to see the surface of these blocks of stone covered with beautiful impressions of what seemed to be leaves; how they got there I could not imagine, but I picked up the broken bits, and even large pieces, and brought them to my repository. (PR 25)

Her description of this period suggests that her unpleasant experience with formal education, instead of turning her against studiousness, made her more determined to pursue studies on her own.

Most observers would have said that there was very little going on at Burntisland, but to Mary Somerville's eye a great deal was going on; over the years, the scientific expertise she acquired expanded her perceptions even more. She "watched the crabs, live shells, jelly-fish, and various marine animals, all of which were objects of curiosity and amusement to me in my lone life." (PR 26) She learned the names of the flora in the surrounding landscape and gathered "ferns, fox-gloves, and primroses, which grew on the mossy banks of a little stream that ran into the sea." (PR 26–27) In her earliest childhood, nature had been her companion; now it had become her open-air classroom and laboratory, a place where she sharpened her skills of observation and attained enormous amounts of first-hand knowledge of plants, animals, and the formations of the landscape. She observed nature with the eye of a painter and the eye of a scientist and stored up a treasury of mem-ories that would become invaluable resources later.

When bad weather kept her indoors, she turned to reading, espe-cially Shakespeare. The epic poems of Ossian, which were set in the Scottish Highlands and contained many vivid descriptions of natural landscapes, were among her favorites and most likely played a for-mative role in developing her mastery of the scientific sublime. In Somerville's youth, especially, Ossian was seen as a master of the sublime and as comparable to Homer. We can get an idea of the tradi-tion that shaped Somerville's and her contemporaries' conceptions of the sublime from Hugh Smith Blair, Professor of Rhetoric and Belles Lettres at Edinburgh and friend of the Somerville family. In *A Critical Dissertation on the Poems of Ossian* (1763), Blair compares Homer and Ossian using terminology that is reminiscent both of Edgeworth's description of the scientific sublime and much of the other commen-tary on Somerville's writing. According to Blair, "Homer's sublimity is accompanied with more impetuosity and fire; Ossian's with more of a solemn and awful grandeur. Homer hurries you along; Ossian elevates, and fixes you in astonishment." (Blair in Ashfield and de Bolla 1998: 209–210) Ossian, according to Blair, raises the mind

> to an uncommon degree of elevation, and wills it with admiration and astonishment. This is the highest effect either of eloquence or poetry: ... Simplicity and conciseness, are never-failing characteristics of the style of a sublime writer. He rests on the majesty of his sentiments, not on the pomp of his expressions. The main secret of being sublime is to say great things in few, and in plain words: for every superfluous deco-ration degrades a sublime idea. (210–211)

These same qualities are the ones often perceived in Somerville and her writing.

When her aunt Janet complained that Somerville read too much and sewed no "more than if she were a man," she "was sent to the village school to learn plain needlework." (PR 28) She was given more freedom to read after she demonstrated her sewing skills. Throughout most of the time she spent at home before she married, she would complete her domestic duties, devote time to developing womanly skills such as needlework and piano playing, and then give whatever time remained to her studies. She had thus already begun the series of negotiations that would make a space in which she could become a scientist, but that space was a long time in coming.

Another valuable strategy she established during her years in her parents' home was disciplined self-education. She obtained a copy of Hester Chapone's *Letters on the Improvement of the Mind* (1772), which offered a detailed description of a prescribed course of reading but suggested that women should not be "remarkable for learning." (139) Like other writers on the subject of young women's education, Chapone had a rather repressive approach to female education. She warned against "the danger of pedantry and presumption in a woman, of her exciting envy in one sex and jealousy in the other, of her exchanging the graces of imagination for the severity and preciseness of a scholar." (139) This was true for "the learned languages" and even more so in the case of "the abstruse sciences." (138–139) Chapone recommended reading history as a way of compensating for lack of personal experience and poetry to cultivate imagination, "the faculty, in which women usually most excel." (141) She recommended the study of nature in terms that seem to have shaped Somerville's own presentation of science: as "a most sublime entertainment" (146) in both the "minute wonders" of the earth and the "stupendous scene" offered by the heavens. (145) In her descriptions of all the wonders the natural world offers to contemplation, Chapone covered all the topics on which Somerville would eventually write and demonstrated all the ways in which the sublime was manifest in them. Chapone defined the purpose of the study of nature as did many others before and after: "to enlarge your mind, and to excite in it the most ardent gratitude and profound adoration towards that great and good Being, who exerts his boundless power in communicating various portions of happiness through all the immense regions of creation." (147)

Chapone's letters offer an interesting insight into the purposes of education for young women as they were conceived at the time. Chapone asserts that

> it is not from want of capacity that so many women are such trifling insipid companions, so ill qualified for the friendship and conversation of a sen-

sible man, or for the task of governing and instructing a family; it is much
oftener from neglect of exercising the talents which they really have. (182)

For women, intellectual improvement is "the sincerest of pleasures; a
pleasure, which would remain when almost every other forsakes them;
which neither fortune nor age can deprive them of, and which would
be a comfort and resource in almost every possible situation of life."
(182–183) Throughout her life, Mary Somerville had the opportunity to
see how great a pleasure and comfort a well-developed mind could be.
The path to influence, according to Chapone, was through "attention
to domestic duties . . . refinement and elegance of manners, and all
those graces and accomplishments." (184) "Endeared to society by
these amiable qualities," she advised, "your influence in it will be more
extensive, and your capacity of being useful proportionably enlarged."
(184) "Studies" were to be subordinate to but supportive of these ends.
Over time, Mary Somerville found that she could attain influence by
bringing together science and domestic virtue. Neither had to be sub-
ordinated to the other if they were both developed to a high enough
degree. She attained enough excellence in both that she did not have
to choose between them.

Somerville was able to follow the course of study Chapone recommended rather easily since her family already owned most of the
books. The small amount of French grammar she had learned at Miss
Primrose's school enabled her, with the aid of a dictionary, to read one
of the recommended works that was in French. With a little help from
a schoolmaster, she learned to use a celestial globe to study the stars,
a terrestrial globe to study geography. It is interesting to consider how
much she resisted formal education, yet thirsted after both knowledge and skill, and to see the ways that the seeds of all the things
she would write about later were sown in these early experiences and
studies. The interests she established as a child stayed with her
throughout her life, though they eventually took a much more sophisticated form.

When she was about thirteen years old, Mary Somerville's world
began to expand geographically, socially, and intellectually. She began to
travel with her mother to Edinburgh for the winter, and these trips seem
to have had a salutary effect on her. She "was sent to a writing school,
where I soon learnt to write a good hand, and studied the common rules
of arithmetic." (PR 35) She received a pianoforte from her uncle, William
Henry Charters, and took a few music lessons. She devoted much time
to practice, developing discipline and probably memory capabilities in
the process. The winters spent in Edinburgh provided her with exposure
and the opportunity to learn new skills. The summers passed at Burn-

tisland provided solitude that was conducive to study and practice. She practiced her piano at least four hours each day and taught herself Greek and Latin. She also began to venture out more in Burntisland and to learn from the people who lived and visited there.

She attended dancing school in Edinburgh, and, in her late teens, began to attend balls, to give supper parties, and to become increasingly at ease in social situations, though she "liked to have some one with whom to enter and sit beside." (PR 62–63) She developed the wide range of social skills and graces that would carry her a long way in the scientific and literary circles of Edinburgh, London, and, eventually, Paris and Rome. She continued to perfect the domestic and artistic skills that would allow her to qualify as thoroughly feminine in the conventional sense. In some ways, these skills and graces would be as important as her mathematical ability in helping her achieve prominence.

She first began to develop an interest in mathematics and science in her early teens. Her first encounters with mathematics took the form of tantalizing yet ultimately frustrating glimpses of incomprehensible code. The story of her efforts to obtain a copy of Euclid reflects her fascination with science, the strength of her memory and powers of concentration, her persistence, and the kinds of difficulties she faced. Given the "fashionableness" of mathematics for ladies in the late eighteenth century, it is not surprising that Mary Somerville was first exposed to higher mathematics through an illustrated ladies' monthly magazine.

> On turning the page I was surprised to see strange looking lines mixed with letters, chiefly X'es and Y's, and asked; "What is that?" "Oh," said Miss Ogilvie, "it is a kind of arithmetic: they call it Algebra; but I can tell you nothing about it." (PR 46–47)

She searched the family library looking for something that would help her understand "what was meant by Algebra." (PR 47) She obtained a copy of Robertson's *Navigation*, thinking that it would enlighten her, but it did not.

> I flattered myself that I had got precisely what I wanted; but I soon found that I was mistaken. I perceived, however, that astronomy did not consist in star-gazing, and as I persevered in studying the book for a time, I certainly got a dim view of several subjects which were useful to me afterwards. Unfortunately, not one of our acquaintances or relations knew anything of science or natural history; nor, had they done so, should I have had the courage to ask any of them a question, for I should have been laughed at. I was often very sad and forlorn; not a hand held out to help me. (PR 47)

She first learned about Euclid in remarks overheard as her painting teacher, Alexander Nasmyth, discussed perspective with two of
her acquaintances. He told them to "study Euclid's Elements of
Geometry; the foundation not only of perspective, but of astronomy
and all mechanical science." (PR 49) "Here," she tells us, "in the
most unexpected manner, I got the information I wanted, for I saw at
once that it would help me to understand some parts of Robertson's
Navigation." (PR 49) But other obstacles remained: she still did not
know enough about algebra to proceed and "as to going to a bookseller
and asking for Euclid the thing was impossible!" (PR 49–50) From
a late twentieth-century viewpoint, the intimidation Somerville felt
about asking basic questions or requesting something so innocuous as a
basic geometry text is hard to imagine. But her fears of being laughed at
or seeming ridiculous were very real and played a significant role in
her life.

Eventually, her younger brother Henry's tutor brought her copies of
Bonnycastle's *Algebra* and Euclid's *Elements of Geometry*. Somerville
was elated: "Now I had got what I so long and earnestly desired." Her
brother's tutor heard her "demonstrate a few problems in the first book
of 'Euclid,' and then I continued the study alone with courage and
assiduity, knowing I was on the right road." (PR 53) Unfortunately,
there were other barriers. When the servants told Somerville's mother
that she was consuming the family's stock of candles by sitting "up
very late reading Euclid . . . an order was given to take away my candle
as soon as I was in bed." (PR 54) The strategy that she used to manage
this setback reveals one of her most important assets a writer and scientist – a strong memory and well-developed imagination, that is, an
ability to call images and symbols up before the mind's eye and to
examine and work with them.

> I had . . . already gone through the first six books of Euclid, and now I
> was thrown on my memory, which I exercised by beginning at the first
> book, and demonstrating in my mind a certain number of problems
> every night, till I could nearly go through the whole. (PR 54)

There may have been other blessings in disguise in the difficulties
of her early life. She says, for example, that lack of money meant she
could not pursue any subject in depth: "I continued my diversified pursuits . . . and was often deeply depressed at spending so much time to
so little purpose." (PR 72) Nevertheless, maintaining diversified pursuits helped her develop skills in managing her time. Although she
focused on "the main object of my life, which was to prosecute my
studies" (PR 64), circumstances forced her to balance her studies with
other activities.

Over time, she turned this necessity into a productive method of working.

> I soon found that it was in vain to occupy my mind beyond a certain time. I grew tired and did more harm than good; so, if I met with a difficult point, for example, in algebra, instead of poring over it till I was bewildered, I left it, took my work or some amusing book, and resumed it when my mind was fresh. Poetry was my great resource on these occasions, but at a later period I read novels. (PR 65)[2]

If she had concentrated on one pursuit, she may ultimately have had less energy even for that one. Moreover, both her ability to establish herself among the elite of the scientific intelligentsia and her status as a role model depended on her social and womanly skills. In the long run, her success as a scientist was facilitated by the diversity of her knowledge and skills and by the abilities she developed as she overcame the obstacles presented to her.

Given the circumstances she faced, it is not surprising that Somerville began to develop resentment along with the strategies she used to reconcile the demands of family and society with her own desire to pursue a course of studies in mathematics. Her resentment tended to be not of the domestic obligations themselves but of the discouragement and intolerance she encountered in the pursuit of her studies. She often felt "kept down" by circumstances that left her with little to say because she could not speak freely about the things that interested her. Sometime near the beginning of her thirteenth year, she began to be annoyed by the disapproval she encountered and "thought it unjust that women should have been given a desire for knowledge if it were wrong to acquire it." (PR 28)

By her late teens, she had experienced both oppression and freedom in the extreme and often in close proximity to each other. Hearing the Liberal party unjustly abused by her father and uncle, as she put it, "made me a Liberal. From my earliest years my mind revolted against oppression and tyranny, and I resented the injustice in the world in denying all those privileges of education to my sex which were so lavishly bestowed on men." (PR 45–46) Though she accurately labeled herself liberal, she was not a revolutionary. Especially in her approach to women's issues, she might be more accurately described as a quiet contrarian who said relatively little but did much to demonstrate that prejudices against women intellectuals were unfounded and unfair.

---

[2] This comment is reminiscent of Sónya Kovalévsky's remarks about literature and mathematics as quoted at the end of chapter one and demonstrates the ways in which discourses that are assumed to be quite isolated from each other can be mutually enriching.

Of all the womanly skills Somerville developed, painting was likely the one most directly useful in her work as a scientist. Somerville herself apparently saw little connection, aside from the fact that she got her cherished information about Euclid from her painting instructor, the noted landscape painter Alexander Nasmyth, who was reputed to have declared Miss Mary Fairfax to be "the cleverest young lady he ever taught." (PR 49) She liked Nasmyth very much: "Mr. Nasmyth, besides being a good artist, was clever, well-informed, and had a great deal of conversation." (PR 49)

Though Somerville may have not have fully appreciated the connection between her painting and her writing, others did. When John Herschel offered suggestions for improving her manuscripts, he appealed to her experience as a painter. In a letter of February 23, 1830, he referred, for example, to "a sketchiness of outline" in her writing, adding, "as a painter you will understand my meaning, and what is of more consequence, see how it is to be remedied." (PR 169) Whewell clearly understood the parallel between the verbal and visual presentation of perceptions of the natural world. Thus, long before she learned to use words to record her observations of nature, she learned to use paint. She approached nature first as a naïve but avid observer, experiencing it in all dimensions, then as a landscape painter, and eventually as a scientist. She presents nature in all of these modes and dimensions in her scientific writing.

In addition to helping her develop her powers of observation, analysis, and presentation, her painting may also have helped build her confidence. She reported that Nasmyth was pleased with her progress. Hugh Blair asked to see some of her paintings and wrote a long letter praising them. "On the whole," he wrote in 1796, "I am persuaded that your taste and powers of execution in that art are uncommonly great, and that if you go on, you must excel highly, and may go what length you please." (PR 58–59) Blair's praise and her family's lack of money led her to think about the possibility of making her living painting, and she concluded:

> I should have been very proud had I been successful. . . . I was intensely ambitious to excel in something, for I felt in my own breast that women were capable of taking a higher place in creation than that assigned to them in my early days, which was very low. (PR 59–60)

Before leaving the period during which Somerville lived with her parents as an unmarried woman, it may be useful to consider in a little more detail the role that her parents played in her development. Like most parents in most times, they seem to have both inhibited and enabled her. Although she disagreed with them on a number of important points, she admired and loved them both, and no doubt drew knowledge and valuable traits from each of them. With her mother, she shared a profound religious faith and a "strength of expression" in the use of language. From her father, she gained knowledge of and interest in navigation, exploration, and other matters related to the seas.

Mary Somerville's life might have been much different if her father had been home more of the time. Especially in her early years, there was a pattern in which she was left more or less to her own devices while he was away and then reined in to a more organized pattern of education when he returned. His efforts to have her read *The Spectator* and Hume's *History* and his investment in boarding school were short-lived as well as unproductive. On the other hand, she seems to have acquired any number of positive influences and qualities from him. Though both parents loved flowers, her father had the more scientific interest in them. He spent time with her gardening and explaining why certain flowers had to be eliminated from the seed pool so that the clarity and definition of shapes of the others would be preserved. He also understood her feeling of awe in the face of nature: "My father, who was of a romantic disposition, smiled at my enthusiastic admiration of the eagles as they soared above the mountains." (PR 66)

It seems that neither of her parents had any conception of what it was that sparked her interest in mathematics and science. Perhaps they were too blinded by conventional notions about women and intellectual pursuits to see the connection between Somerville's avid interest in nature and her equally avid interest in Euclid, failing to understand that one could support the other and that the two together need not render her a "dreaded" woman of learning. At any rate, one of the fortunate coincidences of her life, especially given her parents' conventional views, is that they gave her space – her father primarily because of his travel with the Navy, her mother by virtue of an indulgent nature. From one angle, their treatment of their daughter might be viewed as benign neglect. Regardless of how we decide to label their behavior, it seems that they contributed significantly and positively to the emotional stability she needed to pursue her own interests.

First Marriage and Widowhood

If there is a potential villain in the story of Mary Somerville, it would be Samuel Greig, but his villainy remains only potential, since we know so little about him. Still, one thing seems clear: she came out of her first marriage a changed person, and, in the five years of widowhood that followed, she was transformed as a woman and a scholar.

There is a striking lack of information about Samuel Greig in *Personal Recollections*. For example, the reader gets no clear idea of why Somerville married Greig in the first place. Their entire courtship is treated in one sentence: "My cousin, Samuel Greig, commissioner of the Russian navy, and Russian consul for Britain, came to pay us a visit, and ultimately became my husband." (PR 73) Remarks made earlier in *Personal Recollections* suggest that she married him out of a sense of despair or resignation. She was twenty-four years old, and it was clear that life with her parents would never provide her with the financial and other resources she needed to pursue her studies in earnest. Her family's financial situation also meant that she had no dowry, which would have limited her success in the marriage market, even though she was widely perceived as attractive and charming.

The little information Somerville provides gives every indication that her life with Greig was rather grim and unhappy, and that he was not a sympathetic spouse:

> My husband had taken me to his bachelor's house in London, which was exceedingly small and ill ventilated. . . . I was alone the whole of the day, so I continued my mathematical and other pursuits, but under great disadvantages; for although my husband did not prevent me from studying, I met with no sympathy whatever from him, as he had a very low opinion of the capacity of my sex, and had neither knowledge of nor interest in science of any kind. (PR 75)

It is interesting to consider why a woman who had vowed to raise the general opinion of women's intellectual achievements would marry someone with such a low opinion of the capacity of the sex, and equally interesting to speculate on why a man of such opinions would have chosen her.

Personal Recollections does reveal that Margaret Fairfax had given her daughter twenty pounds to buy a warm shawl for the winter, as part of her trousseau. Mary Greig had used that money to buy a portrait of her father that had been painted shortly after he distinguished himself at the Battle of Camperdown. In the months following her purchase of the portrait, Samuel Greig took her out for drives in his gig. In an unpublished portion of the manuscript, she adds, "*On these occasions I suffered severely from cold as winter came on having only a small scarf; for although I could ask money for the household, I could not ask it for myself.*" (M I 54) There is no indication that Greig did not have the money, only that she would not have felt comfortable asking him for it.

The only other details *Personal Recollections* provides are these from Martha Somerville: "After three years of married life, my mother [Mary Greig] returned to her father's house in Burntisland, a widow, with two little boys." (PR 77) There is no discussion of Samuel Greig's death and no further discussion of his character or behavior. We might attribute this simply to Somerville's assertion that she wants to avoid gossip and personal matters, except that both Somerville and Martha are quite open and occasionally effusive about her very happy marriage to William Somerville.

There is also something curiously telling about the vehemence and persistence with which both Martha Somerville and others deny that Samuel Greig was supportive of Mary Somerville's intellectual pursuits, as was rumored in some obituaries just after Somerville's death.[3] Martha says in her introduction to *Personal Recollections*,

> Nothing can be more erroneous than the statement, repeated in several obituary notices of my mother, that Mr. Greig (her first husband) aided her in her mathematical and other pursuits. Nearly the contrary was the case. Mr. Greig took no interest in science or literature, and possessed in full the prejudice against learned women which was common at that time. (PR 3)

The refusal to provide more information about Greig could reflect loyalty to William Somerville, but it seems equally likely that the reticence arises from an unwillingness to discuss unpleasant matters, a quality the autobiography demonstrates on a number of other occasions and that is typical of Victorian autobiography. (Shortland and Yeo 1996) It is unlikely that the complete story will ever be known.

In any event, it seems reasonable to conclude that Somerville's married life with Greig was difficult, and that his death was somewhat of a liberating experience for her. She came out of it in poor health, so pale that her face *"was like the back of a silver spoon"* (M I 56) and with a child still so young that he was being nursed. But she also had another son, Woronzow Greig, named for one of Samuel Greig's superiors, who would be a great joy and source of admiration and support for her throughout his life. She also came out of the marriage with some financial independence, and, perhaps more importantly, greater

---

[3] The rumor apparently started in the obituary printed in *Nature* on December 5, 1872. The last paragraph of the obituary notice reads: "Dr. William Somerville was his wife's second husband, her first husband having been Captain Greig, a naval officer, fond of mathematics, and who took pleasure in giving his wife instruction in his favourite subject, thus probably giving her mind a bent towards science which has led to important results." (87) The writers for *Nature* tended to interpret Somerville's life in a way that reinforces the notion of science as an exclusively male preserve.

psychological independence. She never stopped to reflect publicly on what she had learned from her marriage, but she behaved differently thereafter. She was no longer as vulnerable to the criticism and social pressure imposed by others, and felt freer to pursue her own interests.

In sum, she seems to have overcome most of the obstacles she had labored under before and during her marriage. The teenager who had "sat up very late reading Euclid" (PR 54) and then rehearsed it from memory in the dark when her candles were taken away now pursued her studies openly and with the benefit of modest financial resources of her own. The girl who had felt "kept down" by family members who made her feel it would be ridiculous to discuss mathematics had found companions who were eager to discuss mathematics and share their knowledge with her. In what might be read as her own personal declaration of independence, she asserts:

> Concealment was no longer possible, nor was it attempted. I was considered eccentric and foolish, and my conduct was highly disapproved of by many, especially by some members of my own family, as will be seen hereafter. They expected me to entertain and keep a gay house for them, and in that they were disappointed. As I was quite independent, I did not care for their criticism. (PR 80)

The "eccentric and foolish" behavior to which she refers includes things such as using her free evening hours (beyond those needed to take care of her children) for studying science and mathematics rather than giving dinner parties. She was able to pursue her studies openly without hiding or suppressing her interest and enthusiasm, as she had often felt compelled to do before, but the intimidation and fear she felt as a girl never entirely left her. She talked freely or openly about scientific subjects only with intimate acquaintances who shared her expertise and interest in the subject. Although she remained thoroughly conventional in virtually every other aspect of her conduct, she was considerably less worried about appearing "eccentric and foolish." (PR 80) Perhaps most significantly, she had a much greater sense of power to control her own destiny.

The growth she underwent in this period is also revealed in the qualities she looked for and admired in William Somerville and in the way she handled the proposals she received before his. One suitor had sent her "a volume of sermons with the page ostentatiously turned down at a sermon on the Duties of a Wife, which were expatiated upon in the most illiberal and narrow-minded language." (PR 88) Mary Fairfax would likely have been wounded or depressed by such an act, or perhaps have simmered in silence. Mary Greig judged the suitor's behavior to be "as impertinent as it was premature; sent back the book and refused the proposal." (PR 88)

Another crucial element in her transformation was her association with the "small society of men" (PR 81) who ran the *Edinburgh Review* and who were part of a large intellectual network of which she eventually became an integral member. These men included Professor John Playfair, who advised her on how to read and master Laplace, and Henry Brougham, who eventually invited her to translate Laplace for the library of the Society for the Diffusion of Useful Knowledge (SDUK). Brougham served as Whig Lord High Chancellor and led an educational reform movement that pioneered several innovations, most of which were concerned with making scientific education more available to the working class.

Somerville's description of her interaction with Playfair gives some idea of the ways in which she became part of this circle and of her well-developed instincts for the social aspects of intellectual life. Playfair "liked female society, and, philosopher as he was, marked attention from the sex obviously flattered him." (PR 82) It is important to remember that Somerville was not only attractive; she was also genuinely interested in mathematics and had the potential to master the discipline.

The third member of that group was Professor William Wallace, professor of mathematics at the University of Edinburgh and one of the most important Scotsmen involved in the movement to modernize mathematics in Britain, which had fallen behind because of "reverence for Newton [which] had prevented men from adopting the 'Calculus' [and] enabled foreign mathematicians to carry astronomical and mechanical science to the highest perfection." (PR 78) Professor Wallace was the editor of a mathematical journal that awarded her a silver medal for solving a prize problem. She went on to become a protégé of his, and he provided her with crucial information and support.

Wallace gave her a list of the books she would need to read to become expert in mathematical and physical science. Professor Wallace's brother, John, studied Laplace with her, and she gained confidence from the fact that she understood it as well as he did. This gave her "courage to persevere." (PR 82) At that point, she probably could not have imagined that she would become not just a beneficiary of the reform movement, but also a key contributor to it. She became much more optimistic in this period, especially after she purchased an "excellent little library" (PR 80) of books on the highest branches of mathematical and astronomical science, nearly all of which were in French, which she had taught herself to read. She says, "I could hardly believe that I possessed such a treasure when I looked back on the day that I first saw the mysterious word 'Algebra.'" (PR 80)

With the acquisition of the library, she had the books she needed to master scientific mathematics. In Wallace, Playfair, and Brougham, she had able guides who could help her master the material. And, of course, she had her own perseverance and ability. She found herself in a situation where her intellectual abilities and interests were taken seriously and she could use her social skills and her femininity to advantage. She had achieved a long-held dream on which she had almost given up. "It taught me," she says, "never to despair." (PR 80)

The events and developments of her widowhood illustrate well the role fortuitous events played in her success. If she had been a widow elsewhere and had not become associated with the men of the *Edinburgh Review* or some similar group, she would not likely have received the encouragement and resources she needed to succeed in mathematical science, nor would she have been as easily able to become part of the scientific network. Her widowhood thus provided the outlines of a figurative room of her own – the financial, social, and psychological space provided by widowhood, and the space within the world of learning opened up in the circle of mathematical reformers led by Wallace, Playfair and Brougham. These conditions were very positive for her development as a scientific intellectual. Her marriage to William Somerville provided the next crucial element.

Marriage to William Somerville

When she reviewed her career, Mary Somerville designated Lord Brougham's invitation to translate Laplace, which she received in 1827, as a crucial turning point in her life, but her marriage to William Somerville in 1812 was an equally important factor in her ability to attain and sustain a prominent place in the world of science. Brougham's letter marked the beginning of her path to widespread public recognition, but excessive attention to it obscures two key issues. First, for the twenty years previous to that invitation, Mary Somerville had been developing the abilities she needed for the task and the connections and reputation that elicited the invitation. The other is that William Somerville performed absolutely crucial functions for her; her life as a woman of science would have been much more limited and less successful without him, and probably not successful at all if she had married another man with the attitudes of Samuel Greig.

As the *New York Herald* of January 4, 1874, put it in a review of *Personal Recollections* entitled "Ideal Old Age," Samuel Greig and William Somerville gave "illustration both to the rule and the exception" with regard to husbands and their support of intellectual wives.

William Somerville would be counted an unusually supportive husband in any era and must have been truly extraordinary by the prevailing standards of 1812, the year in which Mary Greig became Mary Somerville. Her marriage to William would certainly not have been a sufficient condition to set her on a path to public recognition, but it was in some ways a necessary condition.

Somerville's discussion of the reasons William appealed to her suggests that she had a clear idea of what she was looking for in a second husband. What attracted her to William Somerville? Nearly everything – his worldliness, good looks, refined manners, ability to speak English elegantly, fluency in foreign languages, and breadth of learning and interests. But perhaps the most important quality was that he "was emancipated from Scotch prejudices," especially those relating to women, and embraced liberal principles regarding politics, religion, and other topics. He was shunned by some members of his family "on account of his liberal principles, the very circumstance," Somerville later said, "that was an attraction to me." (PR 87)

She was also very fond of his father and mother. His father, the Reverend Dr. Thomas Somerville, had been the first person who had encouraged her studies. Although she had been born in his house, she first remembered meeting him at the age of thirteen. At last, she had found "a friend who approved of my thirst for knowledge." (PR 37) During a visit to the manse at Jedburgh, she and Thomas Somerville took long walks together. He encouraged her to learn Latin by assuring her

> that in ancient times many women – some of them of the highest rank in England – had been very elegant scholars, and that he would read Virgil with me if I would come to his study for an hour or two every morning before breakfast, which I gladly did. (PR 37)

Thomas Somerville not only provided Mary with a positive model of intellectual women; he also helped her find a gentler approach to religion than that presented by the minister at Burntisland, who "was a rigid Calvinist" and whose long, gloomy sermons gave her headaches and made her dislike sermons generally. (PR 33)

She also admired her mother-in-law and aunt, Martha Charters Somerville, who was witty, well read, and a good storyteller. Whereas she felt "kept down" and inhibited in the Edinburgh home of her uncle Henry Charters, she said, "I was never happier in my life than during the months I spent at Jedburgh." (PR 37) Somerville described her mother-in-law as "a very agreeable, but bold, determined person, who was always very kind and sincerely attached to me." (PR 89) This attachment was confirmed in Thomas Somerville's *Life and Times*, when he said that the marriage of Mary Fairfax Greig to their son William "had been the anxious, though secret, desire of my dear wife." (PR 85) Of Mary Somerville, Thomas said, "Her ardent thirst for knowledge, her assiduous application to study, and her eminent proficiency in science and the fine arts, have procured for her a celebrity rarely obtained by any of her sex. But she never displays any pretensions to superiority." (PR 85) It seems that Mary Somerville described Thomas accurately when she said that "he was far in advance of the age in which he lived." (PR 327) This was a characteristic that he passed along to his son William.

William Somerville demonstrated his emancipation from "Scotch prejudices" before their marriage when one of his unmarried sisters, who was younger than Somerville, wrote to Mary Somerville saying she "hoped that I would give up my foolish manner of life and studies, and make a respectable and useful wife to her brother." Mary Somerville was "extremely indignant. My husband was still more so and wrote a severe and angry letter to her; none of the family dared to interfere again." (PR 88)

William Somerville was trained as a surgeon. He entered the army in 1795 as a hospital assistant and received his M.D. at Aberdeen in 1800. His daughter Martha says of him, "Without being very deeply learned on any one special subject, he was generally well informed, and very intelligent." (PR 84) It has often been remarked of him that he had no ambition of his own, a statement that is consistent with the historical record but that overlooks what he had already done by the time he married Mary Fairfax Greig. (They apparently had not known each other until shortly before their engagement.) William Somerville had spent most of his adult life away from England. He "had been present at the taking of the Cape of Good Hope" (PR 86) and had made at least two trips into the interior of South Africa to negotiate with tribes who were attacking Dutch farmers. On the first trip, he had been accompanied by an artist who made drawings of the scenery, people, and animals they encountered. On the second trip, he "was only accompanied by a faithful Hottentot as interpreter." After great difficulty, he had "reached the Orange River, and was the first white man who had ever been in that part of Africa." (PR 87) These adventures were apparently of great enough interest to be published as memoirs, but, for reasons that are unclear, William Somerville never turned his notes into a publishable manuscript.[4]

---

[4] The manuscripts are located in Folders MSWS–1–4 in Dep.c.377 of the Somerville Collection.

William Somerville served as inspector general for hospitals in Scotland and also served in Sicily. Along the way, he had an illegitimate son, James Craig Somerville, whom he named for his commander and friend, General Sir James Craig. This son was educated in Scotland and England along with Mary Somerville's surviving son from her first marriage, Woronzow Greig.[5] Though he was only in his forty-first year when he married Mary Greig, William Somerville could have been accurately described as a man of the world. He had been married and widowed and had already lived an interesting, indeed adventurous life, which may have made him better prepared psychologically and socially to devote primary attention to his wife's career.

William Somerville's own career fortunes fluctuated during his marriage to Mary Somerville. In the early years of their marriage, he was head of the Army medical department in Scotland. His appointment as principal inspector general for England in late 1815 led to a move to London, which had a decidedly positive effect on Mary Somerville's scientific career. But his good fortune did not last long. He was made physician and surgeon to Chelsea hospital in 1819, a position that moved them to much less desirable quarters. After a number of political difficulties and health problems, he retired in 1840, and, from that point on, devoted most of his energies to his wife's career. He seems to have been quite happy sharing scientific and intellectual life with his wife and seemed to take every opportunity of encouraging her development. In the early months of their marriage, for example, he encouraged her to take advantage of her son's tutors to learn more about Greek and botany. Together, they developed an interest in mineralogy, attending lectures and beginning their own mineral collection.

Mary Somerville fully appreciated the difference that her husband's support made: "In those early days I had every difficulty to contend with; now, through the kindness and liberal opinions of my husband, I had every encouragement." (PR 95) Her comments regarding William's conduct during the period of her great celebrity after the publication of *Mechanism* demonstrate how important his support was.

> Our relations and others who had so severely criticized and ridiculed me, astonished at my success, were now loud in my praise. The warmth with which Somerville entered into my success deeply affected me; for not one in ten thousand would have rejoiced as he did, but he was of a generous nature, far above jealousy, and he continued through life to take the kindest interest in all that I did. (PR 176)

---

[5] There is no mention of James Craig Somerville in *Personal Recollections* and only one letter from him in the Somerville collection at Oxford University.

In supporting his wife's scientific career, William Somerville did most of the things that devoted scientific wives have done for their husbands. He offered criticism of her manuscripts, recopied them for her, and searched out the books she needed for her research. As Martha Somerville put it, "No trouble seemed too great which he bestowed upon her; it was a labour of love." (PR 84)

The people who met and wrote about Mary Somerville seemed to be almost uniform in their favorable impressions of her personal demeanor, though sometimes puzzled by its simplicity, especially if they knew her by reputation before meeting her in person. Responses to William Somerville were also generally positive, though less uniformly so than those to his wife. The American astronomer Maria Mitchell, who visited the Somervilles in Florence in 1858, offers this portrait of William Somerville, whom she describes as "an exceedingly tall and very old man . . . in the singular head-dress of a red bandanna turban." The story that follows reflects both Mitchell's gift for storytelling and William Somerville's pride in his wife and desire to see her take center stage.

> He was very proud of his wife, and very desirous of talking about her. . . . While Mrs. Somerville talked, the old gentleman, seated by the fire, busied himself in toasting a slice of bread on a fork. . . . An English lady was present, learned in art, who, with a volubility worthy of an American, rushed into every little opening of Mrs. Somerville's more measured sentences with her remarks upon recent discoveries in *her* specialty. Whenever this occurred, the old man grew fidgety, moved the slice of bread backwards and forwards as if the fire were at fault, and when, at length, the English lady had fairly conquered the ground, and was started on a long sentence, he could bear the eclipse of his idol no longer, but, coming to the sofa where we sat, he testily said, "Mrs. Somerville would rather talk on science than on art." (Mitchell in Kendall 1896: 160–162)

Given the conventions of the day with regard to the dominance of men in public and professional life, as well as in married life, one might wonder how Dr. Somerville's unconventional attitude toward his wife was perceived by others.[6] Although many observers remarked on Dr. Somerville's support for his wife, none seems to have found the role he took either inappropriate or unnatural. That may have been in part because Mary Somerville otherwise assumed such a traditional role; for

---

[6] As Peterson (1989) indicates, it was not unusual for Victorian couples of the middle class to share intellectual interests. Within the Somervilles' circle, there were numerous couples who shared an interest in science. Although William Somerville was an invaluable intermediary for his wife, the intellectual ownership of her scientific work was entirely Mary Somerville's.

example, she was passive in their financial affairs, where he made many transactions, often unwise ones, without consulting or even informing her. William Somerville also took the lead in the face they presented to the scientific and literary circles in which they traveled. In book reviews and on the title pages of her books, she is often referred to simply as "Mrs. Somerville" rather than as "Mary Somerville." All indications are that he was well respected by the other men of their literary and scientific circle and had many close friends among them.

By serving in the role of gentleman companion and intermediary for Mary, William Somerville did a number of things that a scientific "wife" would not have been able to do. He played an active part in the extensive social life through which she made and maintained her important ties to key figures in the world of science. When Lord Brougham wrote asking Mary Somerville to take on a popular version of Laplace, it was William Somerville to whom he addressed the letter and whose assistance he elicited. Her first experiments published in the *Philosophical Transactions of the Royal Society* were communicated by William Somerville, who had become a fellow of that society in 1817. He was her most important link both to the Royal Society and to the British Association for the Advancement of Science.

The statements that Somerville's publisher, John Murray, submitted at regular intervals were addressed to Dr. Somerville; some are headed "Dr. Somerville's View of the Heavenly Bodies"[7] and others "Dr. Somerville's Connexion of the Sciences." The headings eventually became "Somerville's Physical Geography," and so on. Murray sent the statements to William Somerville during his lifetime, to Woronzow Greig after William's death, and to Mary Somerville herself only after Greig died in 1865. William acted as the manager for her career, keeping track of how her work was received, of how her works were selling, and of the profits being derived from them. This last item was particularly important, since her books brought in much needed income for the family. His support gave her space to do what she did best. His experience and assistance, particularly his facility in speaking foreign languages, were also invaluable when they traveled abroad.

At the end of their marriage of nearly fifty years, Mary Somerville reflected on "the sympathy, affection, and confidence, which always existed between us." (PR 326) Those qualities had been noticed by many observers, including Frances Cobbe, who described the Somervilles as "the most beautiful instance of united old age. His love and pride in you, breaking out as it did at every instant when you

---

[7] This was the original title given to *Mechanism of the Heavens.*

happened to be absent, gives me the measure of what his loss must be to your warm heart." (PR 327)

Mary Somerville's problem was not one of finding a man who was willing to marry a woman who was already great – it was a matter of finding a husband who would facilitate greatness. None of the analysis here should be taken to suggest that he was her mentor or that she relied on him for intellectual guidance. Mary Somerville's marriage with William and the enabling effect it had on her career illustrate one of the primary conclusions of Pnina G. Abir-Am and Dorinda Outram's study of women scientists: It is very difficult, if not impossible, for women in conventional marriages with conventional spouses to succeed in science. William Somerville went beyond the husbands described in *Uneasy Careers and Intimate Lives* (1987) in his willingness to subordinate his career to his wife's. Although their middle-class status meant that neither he nor Somerville devoted a great deal of attention to the kinds of housekeeping chores managed by two-career couples today, he did offer other kinds of practical and moral support. Perhaps most significantly, he insulated her from many of the pressures imposed by conservative ideas and helped her transcend the limits imposed by social conventions. Without him, she would have found it difficult to become part of mainstream science or to be recognized for her accomplishments.

John Herschel: A Dear Friend and Valued Critic

John Herschel was the leading astronomer of the nineteenth century and, in the estimation of many, the leading man of science of his day. Although Mary Somerville had connections with many distinguished men of science, her connection to Herschel was one of the longest, closest, and most important.[8] Next to William Somerville, Herschel was the person who played the largest role in supporting her scientific career. They first met in 1816 at the home of Herschel's famous father, William Herschel, who was also a distinguished astronomer. John Herschel was twelve years younger than Mary Somerville. Although she described him as "quite a youth" (PR 105) at their first meeting, they soon came to view each other as contemporaries and colleagues. The correspondence between Herschel and Somerville reflects much affection and mutual respect, and the friendship extended to

---

[8] As Patterson establishes, Somerville's next most important scientific contact was Michael Faraday. (1983: 134) He will be discussed later in the context of Somerville's network of collaborators.

their spouses and families. Somerville was godmother to the Herschels' daughter Matilda Rose (known as Rosa), and the families sometimes spent holidays together. John Herschel had a warm and close relationship with William Somerville as did Mary Somerville with Margaret Herschel.

Herschel and Somerville corresponded extensively on matters related to the substance of science, and it was from him that she learned first-hand of many of his discoveries, new information that would eventually make its way into her books. In a typical letter, written on March 14, 1844, Herschel wrote to Somerville regarding his work in progress – a "monograph of the *principal* Southern Nebulae, the object of which is to put on record every ascertainable particular of their actual appearance and the stars visible in them, so as to satisfy future observers whether *new stars* have appeared, or changes taken place in the nebulosity." (PR 266) Later in the letter, he inquired about "the *real* powers and merits of De Vico's great refractor at the Collegio Romano" in Rome, where Somerville was then residing. Herschel found De Vico's own reports on the refractor "to have not a little of the extra-marvellous in them." Herschel knew that a woman would not be admitted to the Collegio Romano, so he asked, "Has Somerville ever looked through it? On his report I know I could quite rely." (PR 267)

Herschel's letters and published remarks make it clear that he took Mary Somerville seriously as a scientist. In 1845, he communicated her experiments on the effect of the solar spectrum on the juices of plants to the *Philosophical Transactions of the Royal Society*. Referring to them as "elegant experiments" of "the highest interest" and urging her to continue them, he says he is

> sure that they will lead to a vast field of curious and beautiful research; and as you have already once contributed to the Society, (on a subject connected with the spectrum and the sunbeam) . . . it will be a great matter of congratulation to us all to know that these subjects continue to engage your attention. (PR 278–280)

Both her relationship with Herschel and the experimental work he describes in the passage above provide insight into the factors that made Somerville a successful scientist. Her 1845 paper was important, as was her earlier published experimental work in the *Philosophical Transactions*, for establishing her credibility and capability as a scientist with other members of the scientific community. She was then able to draw on this credibility and experience, in reporting on the experimental work of others.

Herschel's most important assistance in her career was probably some of his earliest, as it was connected with the advice he gave regarding *Mechanism of the Heavens*. His responses to drafts of that work show him to be a frank but supportive critic, able to judge from the expert's point of view but also able to appreciate the problems the less expert reader faces. Being one of the leading critics of the day himself, he had a good sense of who the critics were and what their objections were likely to be. In a letter discussing the "Preliminary Dissertation" to *Mechanism* and written on February 23, 1830, he suggested doubling the space given to preliminaries:

> I cannot recommend too much clearness, fulness, and order in the *exposé* of the principles. . . . Your familiarity with the result and formulæ has led you into what is extremely natural in such a case – a somewhat hasty passing over what, to a beginner, would prove insuperable difficulties. (PR 168–169)

He also recommends eliminating needless complications and presenting principles in general terms before going into particulars – all sound principles of the presentation of technical material. His remarks are presented with deference and respect, not from the point of view of an expert leading a novice along, but in the manner of one expert communicating with another regarding matters of strategy in the presentation of difficult material.

Throughout their long association, which lasted until his death, he served as an honest, demanding, and constructive critic of her work, and she relied heavily on his opinions. Despite the enormous success of her work, she never got over a lingering insecurity about her lack of formal training. Herschel's feedback strengthened her manuscripts, and his opinion was valued highly by publishers. But his input also gave her the kind of confidence in the value of her own work that comes from the thorough and frank review of respected colleagues. He encouraged her to continue her work on *Mechanism* and read and revised passages of numerous editions of *Connexion*. When she threatened to burn her manuscript of *Physical Geography* because she was reluctant to compete with Humboldt, who published his *Kosmos* at the same time, she consulted Herschel, "who advised me by all means to publish it." (PR 286) His support continued well into their old age. When John Murray had reservations about publishing *Molecular and Microscopic Science*, he turned to Herschel for advice and reassurance. When she contemplated writing her autobiography, he encouraged her; when he advised her to wait until after her death to publish it, she followed his advice.

In addition to common interests and common acquaintances, there was a sympathy between Somerville and Herschel that went deeper. The qualities that she so admired in him were ones that she herself possessed and cultivated. After visiting the Herschels in 1844, she wrote on September 4 to her daughter Martha, describing the Herschel family as one where "the highest branches of science were freely discussed, but where the accomplishments and graces of life were [also] cultivated." Of John Herschel, she said, "His view of everything is philosophic, and at the same time highly poetical, in short, he combines every quality that is admirable and excellent with the most charming modesty." (PR 271) At the time of Herschel's death, she added:

> He never presumed upon that superiority of intellect or the great discoveries which made him one of the most illustrious men of the age; but conversed cheerfully and even playfully on any subject, though ever ready to give information on any of the various branches of science to which he so largely contributed, and which to him were a source of constant happiness. (PR 361)

As she describes Herschel, she articulates ideals that they shared and, based on the reports of those who knew them, very largely succeeded in living out.

Mechanism of the Heavens

Of all the works that Mary Somerville undertook, *Mechanism of the Heavens* (1831) was the most important for establishing her place among the elite of the scientific community in England. In her autobiography, she asserts that Lord Brougham's 1827 letter asking her to undertake the project "surprised me beyond expression" (PR 162), yet she had spent much of the previous twenty years equipping herself to undertake such a task.

In 1807, she had begun a course of study in mathematical, astronomical, and mechanical science and had mastered its highest branches. The works she studied included those of Francoeur, LaCroix, Biot, Poisson, Lagrange, Euler, Clairault, Monge, Callet, Newton, and, of course, Laplace. Thus, she knew the work of Laplace and his most important contemporaries and predecessors. Since all but two of the works were written in French, she had added a solid knowledge of French scientific and mathematical terminology to the good reading knowledge of French that she already possessed. Through her residence in London and trips abroad, she had become well known in scientific and intellectual circles both within and outside of Great Britain, and Laplace himself, as mentioned earlier, had recognized her as an enlightened and able student of his works before she became one of his best-known translators. When Arago and Biot came to England, they called on her. Although the conclusions of her 1826 experimental paper,

"On the Magnetizing Power of the More Refrangible Rays," had eventually been disproven (a fact that she herself seems eventually to have recognized), the paper established her as a practitioner, a person actively engaged in scientific research, rather than as a friend, patron, or student of science. (Patterson 1983: 48)

The relatively small amount of her experimental work was complemented by frequent, close contact with researchers who were more heavily engaged in scientific investigation. She observed "the glorious appearance of Jupiter" (PR 135) through John Herschel's twenty-foot telescope. William Wollaston came to her house in Hanover Square to demonstrate his discovery of "seven dark lines crossing the solar spectrum."[9] Wollaston later gave the prism he used in that first demonstration to Mary Somerville. In another case, the Somervilles were returning to their home at about 2 A.M. after an evening of observing the stars through a telescope and, noticing that Dr. Thomas Young's light was on, stopped to call. That evening, Young, who had become famous by deciphering the hieroglyphics on the Rosetta Stone, had just deciphered an Egyptian papyrus found in a mummy case. "He had proved it to be a horoscope of the age of the Ptolemies, and had determined the date from the configuration of the heavens at the time of its construction." (PR 131) The comments following this story suggest that Somerville had a good grasp of the substance of Young's work.

Another story illustrating the benefits Somerville gained from close contact with active researchers is contained in an unpublished portion of her autobiography manuscript. When the Somervilles visited James South and his wife at their home in Camden, South invited her to come to his observatory to take some measurements of distances from the sun. He had been teaching her how to make observations and was apparently curious to see what she could do. Her observations were *"very erroneous as might have been expected, but when I took the mean of several observations it differed but little from that Sir James South had made, and here I learnt practically the importance of taking the mean of approximate quantities."* (M I 123) These observations, she concluded, *"worthless as they were enabled me better to describe what others had done."* (M I 122)

These passages and experiences reveal the factors that enabled her to become a highly respected scientist. Her relatively small amount of experience and demonstrated competence in experimental work,

---

[9] In *Personal Recollections*, Somerville says she believed she "was among the first, if not the very first, to whom he showed these lines, which were the origin of the most wonderful series of cosmical discoveries" (133), but this statement is contradicted by a citation in *Physical Geography*, which indicates that the discovery had been published in the *Philosophical Transactions* in 1802.

combined with the things she learned through close proximity with active researchers, gave her a "feel" for science, a knowledge at an experiential level, that was an important factor in enhancing the quality of her later scientific writing.

By the time she received Brougham's invitation, she had mastered the mathematics and science she needed to do the job, had some direct and considerable vicarious experience of scientific experimentation, had the linguistic skills she needed, and had established a solid reputation as a person "of real science"[10] among the people who would be evaluating her work. Brougham was quite confident that Somerville could succeed in undertaking the project; if Mrs. Somerville would not undertake it, he said in a letter written to William Somerville on March 27, 1827, "none else can, and it must be left undone, though about the most interesting of the whole." (PR 162)

Mary Somerville herself was, characteristically, much less confident. At first, she "thought Lord Brougham must have been mistaken with regard to my acquirements." (PR 162) Since her knowledge was "self-acquired," she concluded that it must be inferior to that of university-educated men. She felt "that it would be the height of presumption to attempt to write on such a subject, or indeed any other." (PR 163) Brougham appealed to her by indicating the kind of impact he thought such a work might have: "In England there are now not twenty people who know this great work, except by name; and not a hundred who know it even by name. My firm belief is that Mrs. Somerville could add two cyphers to each of those figures." (PR 162) He was rather vague as to what, exactly, he expected her to produce.

She immediately saw problems with what he proposed. In her opinion, the *Mécanique Céleste* could never be popularized. Readers needed considerable knowledge of both differential and integral calculus, "and as a preliminary step I should have to prove various problems in physical mechanics and astronomy." (PR 163) She would also have to supply diagrams and figures, which Laplace had not provided but would be essential to the readership Brougham envisioned. Brougham may have recognized that she had a better grasp of what needed to be done than he did. In any event, he and William Somerville prevailed upon her to undertake the task, but there were two conditions: the project would be undertaken in secrecy, and if the manuscript was not successful, it would be burned.

---

[10] In his review of the book, Whewell said that *Connexion* "thus shows us how valuable a boon it is to the mass of readers, when persons of real science, like Mrs. Somerville, condescend to write for the wider public, as in this work she does." (Whewell, 1834: 58)

Having made this commitment, the question remained: how to get it done? She had a husband, two children still at home, a household to run, and limited resources with which to run it. She also had a very active social life, which she and her husband both enjoyed very much and in which they had invested considerable energy and financial resources. It was at parties, lectures, and similar occasions that she maintained her connections to the world of science. Moreover, since she had undertaken the Laplace project in secrecy, she could not announce to her acquaintances that she was occupied with a major project and would be unavailable for social calls. "Frequently," she tells us, "I hid my papers as soon as the bell announced a visitor, lest anyone should discover my secret." (PR 164) The middle-class lifestyle that allowed her the freedom to do intellectual work thus also imposed its own demands and constraints.

Although she organized her domestic affairs to give herself time to write in the mornings, she found herself frequently interrupted. She knew only too well that "a man can always command his time under the plea of business, a woman is not allowed any such excuse." (PR 163–164) In this stage of her life, she very likely drew on the skills she had acquired as a young girl who was forced to fit her mathematical studies into the time left over after her other obligations had been met. She developed her skills in time management and mental concentration to the point that she could "leave a subject and resume it again at once, like putting a mark into a book I might be reading." (PR 164)

Somerville's ability to do very demanding intellectual work in the midst of a busy household depended also to a great extent on what her daughter Martha referred to as "a singular power of abstraction": "When occupied with some difficult problem, or even a train of thought that deeply interested her, she lost all consciousness of what went on around her, and became so entirely absorbed that any amount of talking, or even practising scales . . . went on without in the least disturbing her." (PR 164) On one occasion later in life, Somerville became so absorbed in contemplating an idea that she sat in the front row of an audience entirely unmoved and without any change of expression as an Italian improvisatrice changed her subject abruptly and "poured forth stanza after stanza of the most eloquent panegyric upon her talents and virtues extolling them and her to the skies." (PR 165)

How this story ended we are not told, but it serves well to demonstrate the extent of Somerville's ability to disengage herself from external circumstances when she was engaged in a line of thought. This power of abstraction rendered Somerville "entirely independent of outward circumstances," so she did not have "to isolate herself from the family circle in order to pursue her studies." (PR 165) This ability was an important asset for a woman who seemed to thrive in the environment of her family circle and would have found it very difficult to escape had she desired to do so.

It appears to have taken her nearly three years to finish the book. During that time, she employed another strategy developed during her girlhood: refreshing her mind by engaging in a diversity of pursuits. She had developed early on the practice of changing activities when she got stuck or tired. (PR 65) "I was a considerable time employed in writing this book," she tells us, "but I by no means gave up society, which would neither have suited Somerville nor me. We dined out, went to evening parties, and occasionally to the theatre." (PR 166)

Another strategy that she used through most of her career – both before and after *Mechanism* – was the practice of having other experts read and critique her manuscripts. Of these critics, John Herschel was probably the most important. She also consulted with Michael Faraday and with Charles Babbage, who wrote to William Somerville on September 2, 1830, "I cannot find a title for Mrs. Somerville's book; the few pages I saw of introductory matter I liked much." In a letter to Mary Somerville in November 2, 1830, Babbage said, "I am glad to see how you employ common language to illustrate signs – we analysts are too prone to pure symbols." (Somerville Collection, Folder MSB-1 in Dep.c.369)

Mechanism of the Heavens was published in late 1831 and drew an overwhelmingly positive response. Mary Somerville was nearing her fifty-first birthday at the time of publication. Herschel praised the book publicly in a lengthy article in the *Quarterly Review*, and before the manuscript went to press, he wrote Somerville that he had read it with the greatest pleasure,

> and . . . the highest admiration. Go on thus, and you will leave a memorial of no common kind to posterity; and, what you will value far more than fame, you will have accomplished a most useful work. What a pity that Laplace has not lived to see the illustration of his great work! You will only, I fear, give too strong a stimulus to the study of abstract science by this performance. (PR 167)

The final product, though different from what Brougham originally envisioned, was quite successful. Somerville not only translated but also contextualized, elucidated, and interpreted Laplace's work. Because *Mechanism* exceeded the length limits of the publications of the Society for the Diffusion of Useful Knowledge and was not really suited for that readership, Somerville had to seek out another publisher. With Herschel's assistance, she arranged for John Murray to publish the book. He remained her publisher throughout her career.

In addition to the praise heaped on the author of *Mechanism of the Heavens*, there were other honors. The Royal Society voted unanimously to have her bust placed in their Great Hall. She was elected to honorary membership in the Royal Astronomical Society along with Caroline Herschel (aunt of John and sister of William). She also became an honorary member of the Royal Academy in Dublin, the Bristol Philosophical Institution, and the Société de Physique et d'Histoire Naturelle of Geneva.

Perhaps the most substantial recognition she received was a pension on the civil list, recommended to the king by Robert Peel in 1835. Originally £200 per year, it was later raised to £300. Peel said that his objective in giving her the pension was "to encourage others to follow the bright example which you have set, and to prove that great scientific attainments are recognized among public claims." (PR 177) As recognition of her "eminence in science and literature" (PR 177), the pension was both a personal victory and a much-needed financial boost for the Somerville family. But Herschel and others involved in the reform of mathematics and science saw it as a victory for their cause as well. Somerville's was one of the first government pensions given for work in science and thus constituted a step in the professionalization and institutionalization of science, both of which would make it easier for other women to attain the recognition Somerville enjoyed.

Although the high quality of the book was broadly acknowledged, its usefulness was perhaps most clearly recognized by those who sought to modernize mathematics in Great Britain. The group included Playfair, Brougham, Wallace, DeMorgan and Ivory, all mentioned by Somerville, and others, many of whom were associated with Cambridge University, including John Herschel, William Whewell, George Peacock, and Charles Babbage. In addition to his sonnet discussed earlier, Whewell wrote a letter to William Somerville on November 2, 1831, in which he referred to the book "as one of the most remarkable which our age has produced, which would be valuable from anyone, and which derives a peculiar interest from its writer." He concluded that "Mrs. Somerville shows herself in the field which we mathematicians have been labouring in all our lives, and puts us to shame." (PR 170–171)

Professor Peacock's response of February 14, 1832, provides insight into both the character and the value of her accomplishment.

> I consider it to be a work which will contribute greatly to the extension of the knowledge of physical astronomy, in this country, and of the great

analytical processes which have been employed in such investigations. ... Dr. Whewell and myself have already taken steps to introduce it into the course of our studies at Cambridge, and I have little doubt that it will immediately become an essential work to those of our students who aspire to the highest places in our examinations. (PR 172)

Although Somerville was to receive numerous honors and accolades for *Mechanism*, she considered Peacock's and Whewell's use of her book to be "the highest honour I ever received." (PR 172) Her feeling probably derived in part from her respect for Whewell, whom she called "one of the eminent men of the age for science and literature" (PR 172–173), and Peacock, whom she revered as "a profound mathematician." (PR 173) But it likely also grew out of a sense of pride and gratification at having produced something useful to and valued by a group with respect to whom she had once felt herself an outsider.

To put her achievement in perspective, it is useful to compare the experience of Henry Brougham, who eventually completed his own account of Newton's *Principia* and had attempted to write his own popular account of the *Mécanique Céleste*. On September 28, 1840, Brougham reported to Somerville "that though the Cambridge men admit my analysis of the 'Principia' to be unexceptionable, and to be well calculated for teaching the work, yet, *not being by a Cambridge man*, it cannot be used!" In the same letter, Brougham says that he had "almost abandoned" his account of *Mècanique* "in despair after nearly finishing it; I find so much that cannot be explained elementarily, or anything near it. So that my account to be complete would be nearly as hard as yours, and not 1000th part as good." (PR 237)

The extent to that she had become an insider was revealed in the trip that she and William Somerville subsequently took to Cambridge. They were invited by Whewell, then master of Trinity College, to visit Cambridge, where they were provided with "an apartment in Trinity College itself; an unusual favour where a lady is concerned." (PR 179) She was, in all likelihood, the first woman to have received such a favor.[11] Adam Sedgwick made the arrangements for their visit, and they were entertained by many distinguished people of Cambridge University. The kindness she received and the respect she was shown made a profound impression that stayed with her throughout her life. In *Personal Recollections*, she wrote: "The week we spent in Cambridge,

---

[11] This conjecture is supported by a note appended to the description of the Somerville Library at Girton College, Cambridge, which mentions Somerville's 1834 visit and adds, "On that occasion, Mary Somerville may have been the first woman to occupy a set of rooms at Trinity College."

receiving every honour from the heads of the University, was a period of which I have ever borne a proud and grateful remembrance." (PR 179)

A great deal of the importance of *Mechanism* for Somerville's career lay in its status as elite science. Although Lord Brougham had originally said he could guarantee the sale of 1,500 copies of *Mechanism*, John Murray ultimately decided to print 750. Had he printed more, he wrote William Somerville on March 21, 1833, there would have been no profit at all. As it was, though there was little profit, Murray was very pleased to have had "the honour of being the publisher of the work of so extraordinary a person." (Somerville Collection, Folder MSBUS-3 in Dep.c.373) Of the 750 copies printed, 77 were given away, most as the gift of the author. Eventually, nearly all were sold, although Murray's statements related to the matter are somewhat unclear. (No second edition was undertaken.) These numbers illustrate that the prestige of *Mechanism* did not arise from its having been a best-seller. *On the Connexion of the Physical Sciences* and *Physical Geography* each sold in excess of 15,000 copies. (Patterson 1983: 194) As a writer for the *Pall Mall Gazette* put it on February 23, 1874, Somerville had been the only woman of the age to achieve a high reputation in mathematics. People took her reputation "upon trust from men of science." The overwhelming majority of readers had little choice in such matters, since they lacked the mathematical and scientific expertise required to make their own independent judgments. Like the great men of science before and after, Somerville made her reputation in elite science – by producing a work whose nature and value could be appreciated by few and which could have been successfully completed by far fewer, but whose ultimate significance was seen as far-reaching by those who were in a position to confer status.

Maintaining Productivity and Prestige

Mary Somerville spent little if any time resting on her laurels. She began work on her next book, *On the Connexion of the Physical Sciences*, almost immediately. She says she began it because "I was thrown out of work, and now that I had got into the habit of writing I did not know what to make of my spare time" (PR 178), but it seems likely that a need to make money was her primary motivation for writing.[12] Work on *Connexion* started at Chelsea and continued in Paris. (She sent the proof sheets

---

[12] Patterson (1983) puts great emphasis on the financial pressures that led Somerville to write.

through the British embassy.) The celebrity that came with *Mechanism* drew even more visitors and invitations than she had received on her first visits to Europe. She received notable scientists and mathematicians, and exchanged copies of manuscripts and published works. She renewed old contacts and made new ones. After the family's move to Italy in 1838, circumstances made it increasingly difficult for her to obtain books and other written sources for her research, and her interaction with scientific practitioners diminished. But fame also brought people to her and opened doors to institutions that would have been closed to many others, both male and female. There were fewer visitors, but there was also greater intimacy. She had established a discipline that kept her productive, and social connections that kept her active, informed, and intellectually stimulated. She painted more than she had in England and found both the natural and constructed features of the Italian landscape aesthetically rich.

She borrowed books from the Grand Duke's private library at the Pitti Palace, used Professor Amici's microscopes at Florence, and gained access to the library and papers of the East India Company. The research she had undertaken at Chelsea had been simple and required no exotic equipment or materials. The same was true of the work she did in Italy. Researchers – those she knew personally and some she did not know – sent her books and reports of their research. She also had books and other publications sent from England, and corresponded with colleagues in several countries, including John Herschel and Michael Faraday in England. When Henry Brougham visited Rome in 1841, he read over the draft of *Physical Geography* on which Somerville was working at the time and offered advice on how to revise it. She published her second experimental paper through John Herschel in 1846. (PR 278–280) When foreign researchers, such as the American astronomer Maria Mitchell visited Italy, they often called on her, and she became acquainted with many Italian scholars.

The degree to which she stayed involved in scientific life and the extent to which she became part of Italian intellectual culture are reflected in the large number of honors she received after she left England. She was elected to membership or associate membership in the Academy of Natural Science at Florence; the College of Risurgenti; the Imperial and Royal Academy of Science, Literature, and Art at Arezzo; the Italian Society of Natural History; and the Accademia Pontoniana. Near the end of her life, she was awarded the Victoria Medal of the Royal Geographical Society, and in 1870 she received the first gold medal of the Geographical Society at Florence, to which she was also elected an Honorary Associate. She was elected to the American Philosophical Society in 1869.

Elizabeth Patterson's assessment that "the wonder is that Mary Somerville did any science at all after leaving England" (Patterson 1983: 193) reflects both understandable disappointment over the fact that Somerville was not more active in elite science and a somewhat narrow definition of what it means to "do" science. Her assessment may also exaggerate what Somerville would have accomplished if she had stayed in England and underrate what she accomplished in Italy. Somerville was fifty-eight years old when she left England, still remarkably vigorous mentally but at an age when many people consider curtailing their activities. After she left England, she put out four more editions of *Connexion*, produced six editions of *Physical Geography*, wrote *On Molecular and Microscopic Science* (1869), completed a draft and a revision of her autobiography, and published additional experimental work. As subsequent chapters will demonstrate, both *Connexion* and *Physical Geography* were important and influential books.

This is considerably more work than many scientists do after their fifty-fifth birthdays and more output than many generate in a lifetime. She did what she could where she was, and there is no guarantee that she would have been able to do a great deal more in England. The members of her network were getting older, and Somerville outlived most of them. Furthermore, science had become more professionalized and would continue to do so. Even if Somerville had stayed in England, she still might not have been in the mainstream of science.

An assessment very different from and much more positive than Patterson's is offered by Maria Mitchell after her 1858 visit: "Mrs. Somerville, at the age of seventy-seven, was interested in every new improvement, hopeful, cheery, and happy. Her society was sought by the most cultivated people in the world." (Mitchell in Kendall 1896: 163)

Interpreting the Story

Mary Fairfax became the famous Mrs. Somerville as a result of a complex combination of qualities, connections, and accomplishments. Her own talents, effort, and perseverance were the foundation of her success, but they would likely not have paid off as richly as they did if she had not made her way into an extraordinary network of support. Fortuitous events also played an important role in her story. If she had come into full possession of her powers twenty-five years earlier or later, she would likely not have had the same opportunities or recognition. Her association with the members of the Edinburgh circle and the rapport she established with John Herschel are examples of ties that were not sufficient in themselves to make her famous, but that were crucial in her career.

Like the experiences of her formative years, gender was both an asset and a hindrance in the process that transformed Mary Fairfax into the famous Mrs. Somerville. Her attractiveness and feminine social graces made her a welcome addition to the social-intellectual circles of the scientific elite, as well as a charming companion to those who aided her in her scientific pursuits. Being a woman gave her a certain moral cachet and made it easier for her to serve as a symbol of the ennobling power of science. In his review of *Mechanism*, for example, John Herschel had focused on these qualities and potentials:

> Meaner and more selfish motives, may lead a man to toil in the pursuit of science . . . but we can conceive no motive, save immediate enjoyment . . . which can induce a woman, especially an elegant and accomplished one, to undergo the severe and arduous mental exertion indispensable for the acquisition of a really profound knowledge of the higher analysis and its abstruser applications. (Herschel 1832: 551)

Somerville's adherence to the conventions of womanly behavior constituted a living demonstration of the compatibility of progressive science with conservative social and religious values.

Her gender was, thus, a component of the influence she exerted, but it perhaps entered into her life most powerfully in the restrictions it placed on what she could not do, on the avenues and opportunities closed off because she was a woman. Though she was fully admitted to the private inner sanctum of science, the drawing rooms and country houses of Babbage, Herschel, Brougham, and Arago, and the Lodge at Trinity College, Cambridge; though her bust took a place of honor in the Great Hall of the Royal Society at London, she could never have been a fellow of the Royal Society, attended Cambridge or Oxford, or focused her entire life on science.

Perhaps more significantly, she had none of the educational advantages that men such as Herschel enjoyed from a very young age, nor the freedom to devote most waking hours to science that was granted to Babbage, Faraday, and most of her peers. Given the circumstances in which she worked, her accomplishments and the status she achieved were remarkable. In the particular historical moment she occupied, her gender had a much greater impact on what she was allowed to produce and how she spent her time than on how her work was evaluated once she produced it. Her biggest challenge was creating a space where she could do what she did so well.

Her story reveals a number of commonalties and differences with respect to other successful women scientists. She benefited greatly from

both privileged class origin and from having a supportive spouse.
Although William Somerville was a fellow of the Royal Society, Mary
was indisputably the more productive and accomplished in science of
the two. The recognition of this fact by all in the scientific community
who knew the couple, and by William Somerville himself, likely made
it much easier for Somerville to obtain and retain credit for the work
she did. She had no scientist father or brother, but she did have uncommonly supportive in-laws, who heartily approved of her scientific
pursuits. She was necessarily self-educated but benefited greatly from
what Hilary Rose has labeled the "relatively permeable boundary
lines" and "private cultured spaces" that existed in science at that time.
(1994: 99–100) She made a virtue of the need to earn money. That, too,
helped her gain space.

The story of Mary Somerville supports the generalization historian
Margaret Rossiter offers in *Uneasy Careers and Intimate Lives:* " Through
ingenuity, careful planning, and a willingness to defy social convention, some women have devised ways to combine intensive work in
science with time for family duties and pleasures . . . 'science' and
'family,' both broadly defined, have been closely connected for at least
two centuries." (xi–xii)

Like that of many of the women Rossiter was thinking of,
Somerville's path to success was neither easy nor direct. By tracing
that path, we get a clearer view of the many factors that must come
together for anyone – female or male – to reach a high level of success
in science. We also see the role played by both controllable and uncon-
trollable events. Perhaps most importantly, we see how many different
kinds of ingenuity are involved and the necessity of flexibility and
persistence.

The Importance of the Network

Of all the ingenuity Mary Somerville showed, perhaps the most significant was her ability to engage many of the leading men of science
as collaborators in the development of her career. The network of associates that she established beginning during her widowhood and con-
tinuing through her residence in London and trips abroad reads like
a veritable "Who's Who" of the intellectual world: Playfair, Nasmyth,
Blair, Wallace, Wollaston, Young, Kater, Buckland, South, Brougham,
Whewell, Herschel, Sedgwick, Babbage, Arago, Macaulay, Scott, Mac-
Intosh, Biot, Napier, Byron, Laplace, de Candolle, Prevost, Bouvard,
Poisson, de la Rive, Humboldt, Cuvier, Gay-Lussac, Pentland, Marcet,

Quetelet, Schlegel, Edgeworth, Lafayette – the list goes on.[13] Many of the leading cultivators invested in her career in the sense that they contributed to its advancement and identified with its success.[14] This subtle but highly effective strategy meant that she was able to judge at very close range which forms of resistance would be accepted and which would not, and made it unnecessary for her to ever appear to have "demanded" recognition.

One of the central features of her collaborative network was the opportunity it provided to gain input from various experts during the preparation of her manuscripts. The most important of these colleagues were John Herschel, of whom much has been said already, and Michael Faraday, who played less of a role in advising her on career strategy but was second only to Herschel in his importance as a source of scientific expertise. (Patterson 1983: 134) Her interaction with these men and others helped assure the accuracy and acceptability of what she wrote and also helped her to be sure that she was appropriately articulating the collective mind with regard to important new issues and developments in science. She was, to use a modern metaphor, fully integrated and highly interconnected. One of the greatest assets, then, of the space Somerville created for herself was that it was ideally positioned both to facilitate achievement and to garner the recognition that achievement deserved.

[13] Patterson (1983) does a particularly good job of portraying the patterns of Somerville's interaction with all of these figures and offers a much more detailed picture of the structure of Somerville's network than that provided in the text here. See especially chapters 4, 5, and 7.

[14] Most but not all of these people were men. In many cases, Somerville's association extended to both members of a couple, and Somerville closely associated with most of the leading female figures of the day both within and outside of science.

3

Science as Exact Calculation and Elevated Meditation

Mechanism of the Heavens *(1831)*, *"Preliminary Dissertation"* (1832), *and* On the Connexion of the Physical Sciences *(1834)*

For my part, I was long in the state of a boa constrictor after a full meal – and am but just recovering the powers of motion. My mind was so distended by the magnitude, the immensity of what you put into it! I can only assure you that you have given me a great deal of pleasure; that you have enlarged my conception of the sublimity of the universe, beyond any ideas I had ever before been enabled to form.

– Maria Edgeworth, 1832

The "full meal" to which Edgeworth refers is the "Preliminary Dissertation" on the *Mechanism of the Heavens*. The terms she uses and the responses she describes seem more appropriate for an epic poem than a scientific treatise. The "Preliminary Dissertation" combined the qualities of both. Although it was written for and originally published as part of *Mechanism*, the "Preliminary Dissertation" was considered sufficiently valuable to be published independently in 1832. It epitomizes Somerville's most important abilities and contributions as a writer and a philosopher and provided the basic structure for one of Somerville's most popular books, *On the Connexion of the Physical Sciences*, which sold over 15,000 copies.[1] As mentioned earlier, *Mechanism* established Somerville's reputation in elite science. Because *Mechanism* and *Connexion* were Mary Somerville's most esoteric and technical writings, they provide the most convincing demonstration of her ability to portray science as both exact calculation and elevated meditation. They also illustrate the intellectual quality that Whewell defined in gendered terms as the peculiar illumination of the female mind, a quality that ultimately reveals itself more through its blurring of gendered traits than its exemplification of female ones.

The *Mécanique Céleste* embodied a sublime picture of the universe that extended to its outermost limits as well as infinitely forward and backward in time. As one reviewer of *Mechanism* summarized it, "All the changes which can take place in the system were explained, and included in formulae, which represent not merely its present state, but its past and future conditions, even to remote ages." (Galloway 1832: 2) In John Herschel's view, "the history of all the complicated movements of our globe" had been "rendered matter of strict calculation" and could be predicted "from the earliest ages of which we have any record, nay, beyond all limits of human tradition, even to the remotest period to which speculation can carry us forward into futurity." (Herschel 1832: 537) *Mécanique Céleste* thus exhibited comprehensiveness and predictive capacity that had previously been unimaginable. But both the exact calculations and the view of the universe embodied in *Mécanique Céleste* could be appreciated only by those who had extensive background in physical astronomy and could follow Laplace's highly sophisticated methods of mathematical analysis. One of Somerville's greatest accomplishments was to make that sublime vision accessible and real to a much larger group of readers than would otherwise have had access to it.

---

[1] The figures on sales come from Patterson, 1983, p. 193. I have used the British spelling "Connexion" when referring to the title of the book. The edition from which I quote is an American edition, so that the word is spelled "connection" in the quotes included here.

Although this vision had a highly poetic quality, it was not the outcome of a flight of fancy. It was expertise based and grew out of an in-depth understanding of physical astronomy (the subject in which Laplace was working), as well as a very thorough grasp of the observations on which his work relied, the analytical methods he used in arriving at his results, and the practical uses and significance of those results. Somerville was able to convey the complex observations and calculations that contributed to the scope and certainty of the view and to the precision with which it could be delineated. Although this view was implicit in Laplace's presentation, it did not emerge spontaneously from *Mécanique Céleste*. It had to be constructed through interpretation and re-presentation. Somerville's portrayal of science drew on a number of sources and traditions for its power and appeal. Not the least of these were the aura and prestige associated with Laplace himself.

Laplace's Completion of "The Great Work"

To understand what Somerville accomplished in translating Laplace, it is helpful to begin by considering what Laplace himself had accomplished. Laplace has sometimes been referred to as the "Newton of France," a label that provides insight into the breadth of his accomplishments, the esteem in which he was held, and his symbolic significance. Although the scope and complexity of his work make it difficult to characterize, he is generally regarded as having consolidated, extended, and completed the work begun by Galileo and Newton. He had, as Herschel put it, successfully terminated the "great work commenced by Newton, and prosecuted by a long succession of illustrious mathematicians." (Herschel 1832: 537) As the man who had synthesized the contributions of his predecessors and added striking discoveries of his own, Laplace basked in inherited glory and enjoyed considerable renown of his own as a national hero and eventually as a secular saint. (Hahn 1967: 4)

Laplace derived and applied methods for calculating the motions of the planets, their satellites, and the stars. He had applied Newtonian gravitation to the entire solar system and solved some of the most significant problems left by Newton by accounting for the observed deviations of the planets. In the process, he discovered deviations that had not yet been discerned through observation. Perhaps most significantly, he had demonstrated how equilibrium was maintained and proved the stability of the solar system by showing that the deviations, or perturbations in the orbits of the heavenly bodies, were small, periodic, and eventually self-correcting. The techniques he developed provided the basis on which all of the physical sciences could be mathematicized. He publicized his results and the methods by which he achieved them in *Mécanique Céleste*, which was published in five volumes between 1799 and 1825.

Laplace's work was remarkable for its breadth, accuracy, and insight. Historian George Basalla characterizes Laplace's achievement by indicating that it was greeted much as Einstein's work was greeted in the twentieth century and "immediately became a touchstone for the recognition of superior intellect and mathematical talent." (1963: 531) In one version of an often told story, there were rumored to be only twenty men in France who could read and understand Laplace, and only half that number in England. Although Somerville's mastery of Laplace was recognized before she began the project, undertaking an English version of Laplace's *Mécanique Céleste* provided an excellent opportunity to increase her visibility and reputation as a person of science. The prestige of *Mécanique Céleste* arose from both the practical and the symbolic importance of the work, that is, from both the tools and information it provided and from what it said about the age and individual who produced it and about human intellectual capability.

The Challenges Involved

There were, nonetheless, enormous challenges involved, and the fame Somerville achieved was proportional to the challenges she overcame in completing the work. The most obvious of these arose from the bulk and complexity of the *Mécanique Céleste*, which gives new meaning to the phrase "weighty work." Its five books are lengthy and full of abstruse formulae. Laplace was noted for the recondite nature of his subject and for being rather difficult to follow, even for experts in his field. Nathaniel Bowditch, an American who also undertook a successful translation of *Mécanique Céleste*, said, "I never come across one of Laplace's *Thus it plainly appears* without feeling sure that I have got hours of hard study before me to fill up a chasm and to find out and show *how* it plainly appears!" (Bowditch in Johnson 1994: 52) Somerville also had to deal with the less obvious but no less significant hazards arising from being a "learned" woman writing on an esoteric mathematical subject. In the end, she managed to avoid the negative behaviors associated with women writers and to develop a style that allowed her to guide the reader and still appear self-effacing. In this as in other instances, her ability to walk a very fine line between resistance and acceptance of feminine norms was one of the keys to her success.

The hazards of being a woman writer in the early nineteenth century are reflected in the remarks made by John Herschel in his review of *Mechanism*. These remarks are particularly telling because they were made by one of her strongest advocates and most faithful supporters. Although he clearly found much to praise in Somerville's work, he also seems to have shared an expectation that "the learning of a lady," which he puts in the same category as "the poems of a minor," would be marked by inappropriate behaviors such as "female vanity or affectation." (1832: 548) According to this view, women authors are inclined to self-consciousness and prone to calling attention to themselves. Less confident women authors apologize, offer self-deprecating criticism, seek the forbearance of the male reader, or otherwise tax the reader's good nature. A confident woman author, on the other hand, is likely to betray "a latent consciousness of superiority to the less-gifted of her sex" or convey the sense that her knowledge and powers are "extraordinary or remarkable." (548) Although he finds none of these flaws in Somerville's work, he seems to share the prejudices that would have led him to negative judgment if such flaws had existed. He also clearly expects her to conform to womanly norms of modesty and humility and to exhibit a manly unwillingness to ask for special consideration. She does not disappoint him.

As in her reports of her experimental work, which exhibited "perfect freedom from all pretension or affected embarrassment," Herschel finds "the same simplicity of character and conduct, the same entire absence of anything like female vanity or affection . . . nothing throughout the work introduced to remind us of its coming from a female hand." (548) This assessment is particularly interesting in light of Whewell's remarks about the peculiar illumination of the female mind. It suggests that peculiar illumination is not "typical feminine" but what might be called transcendent feminine, that is, a feminine that overcomes the flaws and limits usually associated with femininity while retaining the strengths that come from heightened powers of perception and expression.

The absence of "female vanity or affectation" is the first observation Herschel makes about the work itself in the review and may have been an answer to the first objection he thought his readers were likely to put forward: that the work could not be weighed on equal terms with the work of men because the author asked for special allowances or claimed superiority on the basis of her sex. But in Somerville's case, Herschel says, "We are neither called on for allowances, nor do we find any to make; on the contrary, we know not the geometer in this country who might not reasonably congratulate himself on the execution of such a work." (548) By addressing these negative expectations and arguing that Somerville avoids them, Herschel sets up his own argument – that her work should be weighed on the same terms as that of men and that it compares very favorably on that basis. Establishing this basis for evaluating Somerville's work was crucial to the stature she achieved.

Herschel perceives and presents her as confident, "in perfect consciousness of the possession of powers fully adequate to meet every exigency of her arduous undertaking," but also self-effacing: "In the pursuit of her subject, and in the natural and commendable wish to embody her acquired knowledge in an useful and instructive form for others, she seems entirely to have lost sight of herself." (548) In these comments, Herschel seems to be repeating one of the central precepts of technical writing in the twentieth century: the subject at hand and the objective of conveying an accurate message take precedence. The focus should be on the material communicated, not on the author, who should remain very much in the background. Thus, Herschel might seem to be commending Somerville for achieving what contemporary scientific readers would call objectivity in the presentation of material, the kind of professional stance that makes it inappropriate to use the first person "I" or offer experimental results in the form of a personal narrative.

Yet the 1826 experimental paper by Somerville that Herschel praises in the review is cast as a narrative in first person. It is not dominated by the narrator's "I," but there are numerous instances of phrases such as "I used for the subject of experiment" (132), "My next object was" (134), "I was desirous of ascertaining" (136), "I exposed three pieces of the same steel" (137), and "I am induced to believe" (139). Although Somerville is clearly pursuing what we would recognize as an objective method in her experimentation, and she emphasizes the subject of investigation rather than her own perceptions or experiences, she does use the first person. The example of the 1826 paper demonstrates that objectivity as evidenced through the absence of first-person narrative was not yet a standard of technical writing, or at least that it was not interpreted as it is today. Newton, Laplace, Herschel, and Darwin all used the first person in their scientific writing.

It thus seems reasonable to assume that Herschel is not asserting that Somerville has conformed to an established standard of objectivity in presenting technical material. He is, rather, asserting that Somerville's writing does not display the flaws associated with the writing of a woman. His remarks indicate the ways in which her work was subjected to a particular and peculiar kind of scrutiny. The enthusiasm that Herschel and other reviewers showed for her work suggested that she not only managed to avoid the flaws and weaknesses expected of women but also achieved the standards of excellence established for men in science. She carved out a distinctive territory and style of her own, one that was neither distinctly masculine nor prototypically feminine.

The Content and Character of Mechanism of the Heavens

Mechanism stands in an intricate relationship to Laplace's work and is complex in its internal structure. As mentioned earlier, Lord Brougham had envisioned two treatises, one presenting the more popular view, the other giving an idea of the methods by which the results were obtained. The final form that *Mechanism* took reflected this original design to the extent that it consisted of a "Preliminary Dissertation," which introduced the reader to the subject of physical astronomy and placed Laplace's work in a larger context, followed by her rendering of *Mécanique Céleste*. Neither document qualified as "popular" in the sense Brougham had used the term. The "Preliminary Dissertation" and rendering taken together were, nonetheless, considerably shorter and more accessible than Laplace's original had been. In this instance as in others, Somerville's writing blurs the boundary between professional writing and popularization and raises interesting questions about her intended audience.

Mechanism of the Heavens is sometimes referred to as a "popularization" of Laplace, which implies a non-expert readership and reflects both the wide range of meanings attached to the word and the fact that many people who write about *Mechanism* have never actually looked at it. Consideration of this issue reveals that "novice" is a relative rather than an absolute term, especially given the very small numbers of people who were supposed to understand the work at the time Somerville undertook her version of it. *Mechanism* was intended to make the work of Laplace more accessible to those who wished to master the state-of-the-art in mathematical analysis; it was *not* designed for mathematically unsophisticated readers. The "Preliminary Dissertation" does not, as has often been asserted, provide an introduction to the mathematics needed to follow Laplace. The diagrams that Somerville adds are helpful to a mathematically sophisticated reader but would hardly make the subject clear to a novice. As we saw in chapter two, *Mechanism* was elite science, not meant for mass consumption, though the "Preliminary Dissertation" presented an appealing picture of science to non-expert readers and encouraged interest in science.

Mechanism of the Heavens is often described as a translation of Laplace's *Mécanique Céleste*; it is both more and less. It is more because Somerville contextualized and interpreted Laplace's work, included subsequent contributions of others where they were relevant, and substituted simpler methods of demonstration in some cases. It is less because she treats Laplace's text selectively. One reviewer discerned enough differences and originality in Somerville's rendering to be led to wonder "whether an original work was contemplated, or merely an abridgment of that of Laplace." (Galloway 1832: 4) He concluded that, although the primary intent was to demonstrate Laplace's results, the material was "rendered more simple and perspicuous" where improvement was possible "and advantage [was] taken of recent discoveries in analysis to render the processes more comprehensive and uniform." (4) The result is a streamlined work that is more accessible, original, and up to date than a straight translation would have been.

Somerville provides a wide range of services for the reader of *Mechanism*. Not the least of these is the interpretive framework she provides in the "Preliminary Dissertation," a framework that conveys to the reader the sublime view embodied in *Mécanique Céleste* along with a sense of the larger aims of physical astronomy in particular and the scientific enterprise in general. She also helps the reader locate Laplace's contribution in a long series of developments in physical astronomy, so that the reader can better grasp both the kind of problems Laplace deals with and the significance of his results. A key to the nature of Somerville's accomplishment in *Mechanism* and the impact it had on readers is provided by the working title for the book, which was *View of the Heavenly Bodies*. Though *Mechanism of the Heavens* is an apt description for Laplace's accomplishment, *View of the Heavenly Bodies* more accurately describes what Somerville provided for her readers.

Mechanism is best understood as a rendering of the *Mécanique Céleste*. In this context, "rendering" is intended to convey the notion of translating from one language to another and to connect her work with performance and visual presentation. "Rendering" also refers to interpreting by performing and causing an entity to acquire more of a particular quality, both of which capture the ways in which Somerville's mathematical demonstrations contributed to the advancement of science. The truths revealed through physical astronomy might not be improved, but, as one reviewer put it, the analytical theory and methods of derivation were "susceptible of indefinite improvement . . . rendering the methods already known more simple and uniform; ample scope will always remain for the exercise of the most inventive talent." (Galloway 1832: 3)

There is yet another reason why the term "rendering" is appropriate. Somerville's text, especially the "Preliminary Dissertation," functions much as an artist's or architect's rendering does. It uses words to provide a picture of the complete conception of the universe in motion as it emerges from Laplace's mathematical presentation. Just as an architect's or artist's rendering takes an abstract schematic or verbal representation and presents it as a finished structure, *Mechanism* takes the formulae, observations, and demonstrations of *Mécanique Céleste* and represents them so that they create in the reader's imagination a view of the final construct – the mechanism of the heavens. She allows the reader not only to imagine vividly what the final construct looks like but also to see how it moves.

Although the "Preliminary Dissertation" was originally created to serve as an interpretive frame for Somerville's account of the *Mécanique Céleste*, Somerville seems to have contemplated separate publication very early in the project. Additionally, the "Preliminary Dissertation" is structurally separated from the main text, which has its own introduction. For these reasons, and because the "Preliminary Dissertation" overlaps so much with *Connexion*, the discussion that follows first considers the main text of *Mechanism* and then discusses the "Preliminary Dissertation" both as an independent publication and as a predecessor to *Connexion*.

The Main Text of Mechanism

As part of Somerville's overall strategy of treating Laplace's work somewhat selectively, the main text of *Mechanism* does not give detailed treatment to all of Laplace's material. (All of his topics are mentioned at least briefly in the "Preliminary Dissertation.") She covers the general mechanical principles of celestial mechanics, planetary and lunar theory, and the motions of Jupiter's satellites. There is no detailed discussion of the tides, comets, the attraction of spheroids, or the precession of the equinoxes – all topics that Laplace included. Herschel speculated that these topics may have been left for a second volume. Another reviewer, Thomas Galloway, writing for the *Edinburgh Review*, offered a more interesting and plausible explanation: physical astronomy did not stand still after Laplace completed his work, in part because his work had so powerfully stimulated the use of mathematical analysis in physical astronomy and other sciences. By the time

Somerville began working on *Mechanism*, some of Laplace's work had been superseded. In Galloway's assessment, "Although the *Mécanique Céleste* must ever continue . . . a monument to the genius of the age in which it was composed, it is already in some respects behind the actual state of science." (Galloway 1832: 3)

Somerville does not justify her decisions regarding what she included and omitted and makes very few statements about the purpose or organization of her work. She says only that she has not limited her account of Laplace "to a detail of results" but has instead endeavored "to explain the methods by which these results are deduced from one general equation of the motion of matter." (Mech 3) She also says that she uses higher mathematics instead of the geometrical demonstration that would likely have been more familiar to English readers because she wishes to "give the spirit of La Place's method" and maintain "the unity of his plan." (Mech 3) Although she stops for some internal summaries, she offers almost no other explanation for the strategies she chooses.

We may infer that her decisions were governed not only by a desire to be up to date, but also by an inclination to focus on the most essential concepts and the examples needed to make their implications clear. Since there are fewer departures from Laplace's methods toward the end of her book, we may also infer that some of her decisions about excluding material were shaped by time constraints. Where other analysts had come up with simplified demonstrations, she substituted them, using those of Pontécoulant on more than one occasion. In some cases, she included original, simplified demonstrations of her own and added diagrams. Although she did not succeed in making the material intelligible to a reader who is unskilled in analysis, she did a number of things to guide the reader, such as breaking up the analysis into individual propositions to achieve clearer emphasis.

The introduction to the main text, which is understandably brief given the existence of the "Preliminary Dissertation," defines physical astronomy and outlines the basic principles involved, emphasizing a theme that emerges often in her writing: complexity comprehended within simplicity by powerful natural laws.

> The infinite varieties of motion in the heavens, and on the earth, obey a few laws, so universal in their application, that they regulate the curve traced by an atom which seems to be the sport of the winds, with as much certainty as the orbits of the planets. (1)

She introduces the reader to gravity, the force that often plays the protagonist in the dramas of nature she unfolds for the reader. In some ways, gravitation resembles omniscience, and a knowledge of its workings greatly expands human intellectual capacity. She describes gravity as the

> original property of matter, by means of which we ascertain the past and anticipate the future . . . the link which connects our planet with remote worlds, and enables us to determine distances, and estimate magnitudes, that might seem to be placed beyond the reach of human faculties. (Mech 1)

She links Laplace to one of the central ideas in her philosophy of science – that powerful laws of nature can be discerned through the observation of apparently trivial phenomena: "To discern and deduce from ordinary and apparently trivial occurrences the universal laws of nature, as Galileo and Newton have done, is a mark of the highest intellectual power." (1) She also echoes another common theme: While astronomy has reached a state of perfection, "a wide range is still left to the industry of future astronomers" (2), and there are many mysteries left to be solved. There is much that we are still likely to discover, and much that will likely remain "beyond our reach . . . but instead of being surprised that much is unknown, we have reason to be astonished that the successful daring of man has developed so much." (Mech 2–3) These themes appear often in Somerville's writing, and they form an important part of the rhetoric of science she developed.

Mechanism of the Heavens follows Laplace's organization, somewhat abbreviated and modified. *Mécanique Céleste* consisted of five books, while *Mechanism* has only four in addition to a brief introduction. Book one explains "the laws by which force acts on matter" (150), including the general equation of motion, the uniform and variable motion of bodies and systems of bodies, and the motion and equilibrium of fluids. Book two compares "those laws with the actual motions of the heavenly bodies" (150) and shows how the law of universal gravitation was deduced from observation. It covers the differential equations of mutually attractive bodies, the elliptical motions of the planets and the moon, perturbations of the planets, and the method for constructing and correcting astronomical tables. Book three deals with lunar theory, including lunar inequalities and their investigation through analysis, the numerical values of the moon's coordinates, the inequalities caused by the nonspherical shape of the earth, the action of the planets on the moon, and Newton's lunar theory. Book four deals with the theory of Jupiter's satellites, their perturbations and inequalities, the numerical values of the perturbations, and the eclipses of Jupiter's satellites. The main text ends abruptly after a brief and rather undeveloped discussion of the satellites of Saturn and Uranus. There is no closure of any kind, a fact which supports the conjecture that her strategy may have been somewhat governed by the time she had available. This abruptness is less problematic than it might seem because of the cohesiveness provided by the conceptual framework developed in the "Preliminary Dissertation."

Somerville's Roles with Respect to the Material and the Reader

Somerville takes several roles with respect to the reader and the French text of *Mécanique Céleste*, and shifts rather abruptly and without warning from one to another. Sometimes she plays the role most often associated with translation, that of the transparent window conveying a straightforward English version of Laplace's work. In these cases, she retains the explanatory strategy, structure, and methods of demonstration used by Laplace. She plays this role with increasingly frequency toward the end of *Mechanism*.

One of her most common roles is that of guide or intermediary who establishes some distance from the material, distinguishes herself from Laplace and explains the strategy Laplace is using, helps the reader follow Laplace's analysis, resolves apparent contradictions, or offers encouragement to the reader. At the end of book one, she stops to encourage the reader, recognizing how "arduous" and "tedious" the study of the material is. She emphasizes the prize to be had at the end of the effort – "the contemplation of the most sublime works of the Creator" – and she uses the difficulty the reader has been experiencing to highlight Laplace's genius and give us a better appreciation for what he accomplished. For example, she indicates that the values of the unknown quantities in the general equation of motion can be precisely determined in only a few cases. "These," she tells the reader, "La Place has selected with profound judgment, and employed with the greatest dexterity." (144)

She also acts as guide at the end of book two, chapter two. Having gone through the demonstration of the law of universal gravitation, she steps back and explains what Laplace is doing:

> Having thus proved from Kepler's laws, that the celestial bodies attract each other directly as their masses, and inversely as the square of the distance, La Place inverts the problem, and assuming the law of gravitation to be that of nature, he determines the motions of the planets by the general theorem in article 144, and compares the results with observation. (Mech 169)

She then explains how the motions were calculated, addresses logical but incorrect conclusions that readers might reach, and presents the implications of the principles that have been established. All of this leads to the conclusion that

> this conformity of nature with itself upon the earth, and in the immensity of the heavens, shows, in a striking manner, that the gravitation we observe here on earth is only a particular case of a general law, extending throughout the system. (Mech 168)

This conclusion reinforces the notion of the pervasiveness and power of the law of gravity. Throughout *Mechanism*, she often uses this strategy of following a demonstration of a concept with a discussion of the more general principle to which it relates and an explanation of its practical significance.

Somerville also takes on the role of historian of science or scientific biographer. She uses history to develop the background that gave rise to the problems Laplace solved and to provide a sense of perspective. She does this to develop the conceptual framework related to the accomplishment in question, to relate it to a concrete problem in physical astronomy, or to establish the significance of Laplace's contributions and the genius or ingenuity they reflect.

An example of this occurs at the beginning of book two, chapter one, where she includes a rather lengthy passage on the history of astronomy. She describes the discoveries of Copernicus, the trials and discoveries of Galileo and Tycho Brahe, and the contributions of Kepler, "one of those extraordinary men, who appear from time to time, to bring to light the great laws of nature" (147), all leading up to Newton, who "by his grand and comprehensive views, combined the whole, and connected the most distant parts of the solar system by one universal principle." (148) She describes the course of Newton's reasoning and his significant discoveries. She traces a line of "brilliant discovery" through Newton's *Principia*, "a work which has been the admiration of mankind, and which will continue to be so while science is cultivated" (149) and establishes Laplace's reverence for Newton, although she says that Laplace "perhaps only yields to Newton in priority of time." (149) After she establishes the esteem in which Laplace held Newton and quotes a letter she received from Laplace in which he states his faith in the future progress of British mathematics, she leads up to the key problem Laplace solved:

> The reciprocal gravitation of the bodies of the solar system is a cause of great irregularities in their motions; many of which had been explained before the time of La Place, but some of the most important had not been accounted for, and many were not even known to exist.

The author of the *Mécanique Céleste* therefore undertook the arduous task of forming a complete system of physical astronomy, in which the various motions in nature should be deduced from the first principles of mechanics. (150)

She also plays the role of revisionist who substitutes simpler demonstrations, sometimes those of others, sometimes her own. For example, in book two, chapter six, where she discusses the secular inequalities in the elements of the orbits, she indicates that part of the analysis she has presented comes from Laplace and part from Poisson. (258) She also brings the reader up to date on subsequent developments. For example, at the end of chapter six of book two, she takes a step back from the demonstration of the stability of the planetary system to mention a recent paper that refined the hypothesis on which Laplace and Poisson's proof was based. (288) Somerville thus plays many roles as the translator of Laplace, from transparent window to guide, historian, revisionist, critic, and interpreter.

For Somerville and other women both before and after her, translation was one of the most available forms of serious participation in science.[2] Although translations of significant scientific treatises were often very badly needed by the scientific community and required considerable technical expertise, they were usually not considered to be a suitable use of time for an active experimentalist or theorist. This created an opportunity for women, and translation was consistent with the social roles women were expected to take. It seemed compatible with the role of helpmate and offered the satisfaction of doing something useful. Translators typically have not been accorded a large role in the history of science. Nevertheless, close examination of translations reveals the expertise that translators needed, the critical spirit in which they approached the works they translated, the ingenuity they often displayed, the contributions they made, and the recognition they sometimes received for their work.

Much of the difficulty in understanding what translation involves and why it matters arises from the broad spectrum of meanings of "translation," which range from more simply putting a document

---

[2] In fact, it might be argued that translation has been somewhat of a traditional role for women in the history of science. Aphra Behn (1640–1689) translated Fontenelle's *Entretiens sur la pluralité des mondes* into English; Emilie du Châtelet (1706–1749) translated Newton's *Principia* from Latin into French; Claudine Guyton de Morveau (ca. 1770–1820) translated the works of several European scientists into French; Ada Byron Lovelace (1815–1852) translated Luigi Menabrea's treatise on Charles Babbage's *Analytical Engine* from French into English; and both Marie Lavoisier (1758–1836) and Mary Lyell (1808–1873) translated scientific documents for their husbands. (Ogilvie 1986)

in the words of another language, to the sort of rewriting, reworking, critique, and rendering that Somerville carried out in *Mechanism*. Translations often exist not just to make a text available but to make it more accessible and useful. This may require reorganization or clarification, as well as contextualization and commentary. One cannot assume that the content and organization of the original text have been perfected.

The translation of abstruse material like that in *Mécanique Céleste* requires more than a reading knowledge of the original language; it requires a background in the subject being discussed and a knowledge of the technical terminology involved. Furthermore, the nature of the material itself called for creative and inventive effort just to be comprehensible. As one reviewer of *Mechanism* described the endeavor:

> The analytical processes by which the more refined truths of astronomy are reached, are of so abstruse a nature, and so far removed from ordinary apprehension, that they who contribute to render them more easily understood, may justly claim to be regarded as benefactors of science. (Galloway 1832: 3)

The translator often develops an unusually intimate understanding of the work being transcribed, which can put her or him in an excellent position to provide informed commentary, especially if the translator has the kind of wide acquaintance with the literature that Somerville possessed. In her rendering of *Mécanique Céleste*, Somerville operated across the entire spectrum of possibilities for intervening as a translator, from no intervention at all to original input of her own.

One of the central points to recognize is that translation at any point on the spectrum is not an unmediated act; it is always to some extent a re-presentation and re-interpretation, even when the material being translated is scientific or technical and presumably "objective." It is also important to recognize the variety of women's reasons for becoming translators and the many goals they have sought to achieve through their work. Translation, like the dissemination of scientific knowledge generally, has many purposes. Some translations are designed to domesticate science, to make it accessible to those whose interest in science is more social than intellectual. But other forms of translation, like Somerville's, are undertaken in an intellectual and critical spirit and have often been recognized as significant contributions to the advancement of science.[3]

---

[3] For an interesting discussion of women as translators in the eighteenth century, see Paula Findlen, "Translating the New Science: Women and the Circulation of Knowl-

"Preliminary Dissertation": Creating an Interpretive Frame

One of the most difficult decisions Somerville had to make in writing the "Preliminary Dissertation" was how to start – how to orient the reader and put the reader in a position to appreciate and understand the material to come. This was not an easy decision given the nature of the work she was introducing. The prestige of its author was widely acknowledged, but its subject matter was recondite and not well understood. Moreover, the work itself was enormously broad and represented a synthesis of the work of a long line of "illustrious mathematicians," whose works were often equally difficult to understand, though perhaps better known. There was in addition the problem of making the material interesting and relevant, a problem that can be best understood only by those who have actually looked at the work itself and understood its challenges.

Her solution was to present the view of the universe that emerged from the accumulated efforts of Laplace and his predecessors, a view that grasped both the mundane and the sublime. She accomplished all of these objectives in the space of only seventy pages. These seventy pages exemplified the abilities and qualities of her writing that were to be her greatest strengths throughout her career – the capacity to help the reader visualize the view of the natural world that emerged from physical science, an ability to associate that view with elevated feeling and beauty, a capacity to achieve clarity and precision without providing overwhelming amounts of detail, and an ability to connect esoteric theories to problems of everyday life. In the process, she created a powerful interpretive frame for science and articulated the intellectual foundations on which the remainder of her scientific work and writing would be built.

This interpretive frame consisted of two distinct components. The first was a set of ideas and philosophical commitments, a belief system within which science was located and which seemed to preexist science, but that also seemed in some ways to emerge from the study of the natural world through science. In the analysis that follows, these ideas and beliefs will usually be referred to as recurring themes. A

edge in Enlightenment Italy," *Configurations* 2 (1995):167–206. For information on women writers on science in the eighteenth and nineteenth centuries, see Kathryn A. Neeley, "Woman as Mediatrix: Women as Writers on Science and Technology in the Eighteenth and Nineteenth Centuries," *IEEE Transactions on Professional Communication* 35, no. 4 (1992): 208–216.

number of these have already been highlighted in Somerville's work; they are treated in a more consistent way in this chapter. The second component consisted of descriptive patterns, that is, commonly used strategies for presenting explanations of scientific phenomena. Somerville used a number of these in her scientific writing. In the "Preliminary Dissertation" and *On the Connexion of the Physical Sciences*, the two most important descriptive patterns were what has been referred to earlier as the "cosmic platform" and another pattern I will call "tracing the mazes." Because the interpretive frame is best understood within the context of the overall structure of the "Preliminary Dissertation," the discussion that follows begins with the general characteristics of the "Preliminary Dissertation" before going into the details that make up the interpretive frame.

The Overall Structure of the *"Preliminary Dissertation"*

As one of the reviewers of *Mechanism* pointed out, Somerville is not very explicit about what she hopes to accomplish either in the "Preliminary Dissertation" or in *Mechanism* as a whole. She casts the "Preliminary Dissertation" as "a few preliminary observations on the subject which [this work] is intended to investigate, and of the means that have already been adopted with so much success to bring within the reach of our faculties, those truths which might seem to be placed so far beyond them." (iv) The dissertation begins with general observations about scientific method and physical astronomy but does not proceed according to any clearly defined plan or map of the territory to be covered. She moves from topic to topic by a series of local connections, in which each topic seems to be linked somehow either to the one that preceded it or the one that follows, but the overall design of topics is not clear.

She covers all of the topics that Laplace covered in the *Mécanique Céleste*: (1) the basics of the law of gravitation, that is, the simple principles that operate in a multitude of complex situations; (2) rotational motion and all of its consequences; (3) translational motion; and (4) the forces of gravitation as manifest in the tides and other actions on the atmosphere. Near the end of the "Preliminary Dissertation," she deals with a set of miscellaneous topics related in one or more ways to those she has already discussed: the nature of light, sound, and heat; magnetism; comets; and problems related to observing and understanding the stars. Her basic explanatory strategy is to start with the simplest case of a given phenomenon and use that as the basis for explaining more complex phenomena. She describes techniques used to make the problem tractable, such as the approach to the three-body problem. She also uses epitomes, or small-scale examples, of large-scale phenomena.

To clarify the operation of concepts, she often creates hypothetical cases to develop principles or contrast the situation as it really is with the situation as it might be mistaken to be. In every area she covers, she indicates the ways that various developments expand human capability, intellectual grasp, or predictive power and discusses a number of practical uses of the findings and methods of physical astronomy, including the usefulness of the lunar theory in navigation, the use of astronomy for chronology (especially the dating of objects and events based on astronomical data), and the use of knowledge about the earth to furnish standards for weight and measures. Each of these examples relates to one of her most common themes – the idea that there are phenomena in nature that are invariable in the face of the vicissitudes that affect other aspects of human existence and can be used as benchmarks or standards of measure.

Although she does not devote a lot of time in the "Preliminary Dissertation" to Laplace, she mentions him fairly often in order to locate his contributions in relation to earlier discoveries and problems, establishing both a line of discoveries to which his work contributes and the larger framework of human activities, concerns, and problems into which they fit. She thus introduces the reader to the *Mécanique Céleste* by delineating the evolving view of the universe that had developed prior to Laplace and indicating where he had made contributions. She also relates the topics she takes up to activities and concerns that are likely to be familiar to the reader. In a strategy that she duplicates many times, she moves from the most abstract and universal of principles through their many consequences, engaging in an activity that she likens to following a force through a maze.

From the beginning, the "Preliminary Dissertation" attracted very favorable attention. The comments reviewers made about it provide insight into the combination of factors that drew such favorable attention. Thomas Galloway pronounced it a detailed and interesting presentation of

> most of the striking facts which theory and observation have made known respecting the constitution of the universe. This discourse is . . . calculated to give us a very high opinion of the industry and scientific attainments of its author; as it displays a correct and intimate acquaintance, not only with theoretical astronomy, but with the whole range of physical science, and the best and most recent works which treat of it. (1832: 6–7)

John Herschel praised it for its combination of "precision and clarity," factors

> which render this *Preliminary Dissertation* a model of its kind, and a most valuable acquisition to our literature. We have no hesitation in saying, that we consider it by far the best condensed view of the Newtonian philosophy which has yet appeared. . . . an abstract so vivid and judicious as to have all the merit of originality, and such as could have been produced only by one accustomed to large and general views, as well as perfectly familiar with the particulars of the subject. (1832: 550)

She had accurately and concisely conveyed the view of the cosmos provided by physical science. The "Preliminary Dissertation," then, might be characterized as a concise state-of-the-art account of the constitution and operation of the universe as they were understood at the time.

Galloway identified another important strength of the "Preliminary Dissertation" when he said,

> The whole is eminently calculated to inspire a taste for the pleasures and pursuits of science; and to promote a desire to penetrate the recesses of that sublime geometry which presides over the motions, and determines the forms and distances, of the planetary bodies. (1832: 7)

Somerville had added to the accomplishment of making the view clear, the excellence of making it appealing. She had exploited the aesthetic dimensions of the view to make the reader more aware of the aesthetic pleasures of science and the way they were intimately related to the intellectual pleasures of science.

The scientific sublime Mary Somerville evoked through her writing combined religious, aesthetic, and scientific elements into a distinctive style of presenting science. When Maria Edgeworth praised Somerville's ability to evoke the scientific sublime, she focused on the simplicity and restraint of Somerville's writing style, and said she was gratified by the way Somerville trusted the natural interest of her subject and relied upon the "reader's feeling along with" her, rather than using "ornaments of eloquence dressing out a sublime subject." (PR 204) This quality of transparency was probably at least in part Somerville's response to gender norms, but it also served an aesthetic and rhetorical purpose by making her writing more appealing and more persuasive.

Somerville was able to evoke the scientific sublime so deftly because the frame she used set forth overarching themes that became the context or intellectual backdrop for the material that followed. Another reason she was able to achieve a great impact with a relatively restrained style was that she drew on a very rich poetic tradition that associated scientific study of nature with the sublime. One of her greatest accomplishments was that she took this well-established poetic tradition and adapted it to scientific prose. The significance of her accomplishment lay in the extent to which she integrated poetic themes and the presentation of scientific information and in the subtlety with which the effect was achieved.

Colin Ronan has asserted that "Bacon's importance lay in the stimulation his ideas aroused, in his vision of the betterment of the human condition through the application of science" (1983: 373), rather than in the method he proposed. In a similar vein, Somerville's primary contribution derived from the vision of science she presented. In that vision, the pursuit of science was blended harmoniously with religious, moral, and aesthetic values and with the pursuit of practical advantages and applications; she thus combined a persuasive big-picture view of the scientific enterprise with the most current information about developments in physical science. The value of Somerville's writing had its foundation in her own expertise and the access that she had to the researchers who were most closely involved in advancing science. But its influence arose from providing readers with a particular vision or interpretation of science and uniting feeling with knowledge in an intricate weaving of scientific, literary, and religious discourses.

This weaving allowed her readers to perceive patterns of association and unity that could link fact with fact, discovery with discovery, and scientific discourse with other forms. In a review essay on Galileo, William R. Shea commented on Galileo's "outstanding rhetorical gifts and his dialectical skill," factors that are often overshadowed by his other accomplishments and problems. Shea says, "Galileo did not offer his ideas in the nakedness of abstract thought, but clothed them in the colors of feelings, intending not only to inform but also to move and entice to action." (1994: 483–484) Somerville's gift was similar to that of Bacon and Galileo, the capacity to provide a vision of science and associate positive feeling with it. Somerville's vision was not as programmatic as Bacon's or as revolutionary as Galileo's, but it played a significant role in helping to transform scientific facts and theories into a scientifically informed yet culturally grounded world view.

Thus, Somerville's success was greatly facilitated by the interpretive framework in which she located the enterprise of physical science. This framework drew heavily on themes that had already been established by the scientific poets in the eighteenth century as well as on ideals of the intellectual circle in which she participated. It had a number of themes and elements, the most important of which was the idea that science provided a cosmic platform, a view from space where the reader was invited to contemplate the immensity, regularity, intricacy, and beauty of the creation. The platform was intimately connected to life on earth through the law of gravitation. Contrary to what we might expect today, the view from the cosmic platform did not allow the reader or scientist to usurp the position of God. Rather, it humbled the reader by reminding him or her of the immensity of creation and the smallness of both human life and earthly existence in the overall scheme of things. It was designed to inspire humility and admiration for the creator, even as it emphasized the human power and potential embodied in the capacities needed to practice and to advance science. It presented not a cold and mechanistic universe, but a universe alive with color, light, force, and movement and imbued with the goodness and wisdom of the "Almighty Architect" who created it.

In sum, the role of the frame was to help readers visualize the sublime picture *Mécanique Céleste* presented and to interpret it within the framework of beliefs and values of the rising middle class and scientific reformers – science and religion, science and humility, science as elevated meditation and exact calculation, science as the pursuit of goodness and virtue. It also conformed to and promoted an aesthetic that abhorred coldness, colorlessness, and mechanical views, and embraced a vivid, warm, aesthetically satisfying picture.

Herschel expressed this aesthetic as it related to science when he referred to Laplace's findings on the periodic and ultimately stable deviations of the planets as a "noble theorem" that "forms a beautiful and animated comment on the cold and abstract announcement of the general law of gravitation." (Herschel 1832: 538) Somerville's portrayal of a warm, colorful, and animate universe had a deeper significance than might be immediately apparent. Laplace's title, *Celestial Mechanics* or *Celestial Mechanism*, suggested a mechanistic and deterministic view of natural processes. Somerville's interpretive framework incorporated the findings and techniques of Laplace within a belief system and view of nature in which the Creator still had a very important role, though He was now much more the designer than the hands-on manager of the universe.

This belief system incorporates attitudes or stances we take to be contradictory. These include: (1) pleasure in immensity or cosmic scale and in the speed of movement and breadth of perception required to traverse enormous distances; (2) pleasure in perceiving complexity within regularity, in discerning unity or design, in seeing the same patterns manifested in many different contexts, in connections that are meaningful but not obvious; (3) exhilaration at the view enjoyed by a transcendent consciousness; (4) pleasure in mystery (that which is not yet known and may be radically different from what we already know or assume to be true) combined with a glorification of accumulated knowledge; and (5) pleasure in recognizing the similarity between human and divine intelligence combined with a profound humility regarding the small role of man and of Earth in the immensity of the universe. Much of this aesthetic is, as has been demonstrated by William Powell Jones (1966), borrowed from the scientific poets, but it has added dimensions of its own.

The main idea underlying the frame is that science brings together elevated meditation and exact calculation, all qualities of the divine mind, which are duplicated or mirrored in the mind of man. The more exact the calculation becomes, the more elevated the meditation, because the breadth of view and power of prediction also increase. The ideas that form the basic building blocks of the frame are developed at the beginning and end of the "Preliminary Dissertation" and in the introduction and conclusion of *Connexion*. They are illustrated amply but only occasionally referred to in between. Still, once the frame is established, it is easily brought into view and used to evoke the sublime. The frame is made up of the two kinds of components mentioned earlier: themes and descriptive patterns.

The Themes

Most of the themes in the conceptual frame developed by Somerville connect science to religious and aesthetic traditions; some are concerned with philosophy of science – the relation of knowledge to experience and of mathematics to an understanding of the physical universe. The discussion below develops the individual themes as they are presented in the "Preliminary Dissertation." In *Connexion*, there are some minor changes, but the framework and central themes remain essentially unchanged.

First, "design is manifest in every part of creation" (Mech lxvii), and the most important divine design criteria are simplicity and stability. The form that our universe takes is not the only one that would have been possible given the existing physical laws, and "it may be concluded, that gravitation must have been selected by Divine wisdom out of an infinity of other laws, as being the most simple, and that which gives the greatest stability to the celestial motions." (lxviii) These laws are invariable throughout the solar system and "immutable like their Author. Not only the sun and the planets, but the minutest particles in all the varieties of their attractions and repulsions . . . are obedient to permanent laws." (lxix) This constancy does not preclude new discoveries, however: "Perhaps the day may come when even gravitation, no longer regarded as an ultimate principle, may be resolved into a yet more general cause, embracing every law that regulates the material world." (lxix)

The most convincing single proof of this unity is the pervasiveness of gravity, "that extraordinary power, whose effects we have been endeavouring to trace through some of their mazes." (lxviii) Gravity thus becomes a manifestation of the pervasive power and presence of the Creator, and understanding its complexity and powerful behavior is aesthetically, intellectually, and spiritually satisfying.

Second, scientific knowledge arises from but is superior to ordinary experience. New laws are discovered through "experience, which furnishes a knowledge of facts, and the comparison of these facts establishes relations, from which, induction . . . leads us to general laws." (v) Experience can, of course, be misleading, and science is a corrective or supplement to human faculties in those instances. Knowledge thus advances through experience, and through deduction (proof through comparison or analogy) and induction (building up general truths and relations from facts). "By such steps [Newton] was led to the discovery of one of those powers with which the creator has ordained that matter should act reciprocally on matter." (v) Analogy is, thus, not just an explanatory strategy, a way of making unfamiliar ideas comprehensible; it is also an important conceptual tool for advancing science.

Third, the contemplation of nature through science is a fundamentally ennobling activity that has transcendent aims. Specifically, "the contemplation of the works of creation elevates the mind to the admiration of whatever is great and noble [and] to inspire the love of truth, of wisdom, of beauty, especially of goodness, the highest beauty, and of that supreme and eternal mind, which contains all truth and wisdom, all beauty and goodness." (vi) Science is the all-inclusive pursuit of truth through impartial and patient investigation: "Nothing is too great to be attempted, nothing so minute as to be justly disregarded." (vi) Pursued for these aims only, science raises the mind of human beings and prepares us for "those high destinies which are appointed for all those who are capable of them." (vi) As a pursuit that "must ever afford occupation of consummate interest and of elevated meditation" (vi), science thus has moral, aesthetic, and religious as well as intellectual value.

Fourth, science traces unifying principles throughout the universe. This idea is expressed through the notion of "an uninterrupted chain of deduction from the great principle that governs the universe" (v) in which the perception of analogies plays an important role.

Physical astronomy is the science which compares and identifies the laws of motion observed on earth with the motions that take place in the heavens, and which traces . . . the revolutions and rotations of the planets, and the oscillations of the fluids at their surfaces, and which estimates the changes the system has hitherto undergone or may hereafter experience, changes which require millions of years for their accomplishment. (v)

In this scheme, exact calculation progresses to and supports elevated meditation.

Fifth, this ennobling "ascent to the starry firmament" induces humility in human beings rather than arrogance because their expanded understanding of "the immensity of creation" (vii) leads them to see how very small the earth and humanity are compared to the universe and to expect that there is a great deal more in the universe to be discovered, including things we cannot yet imagine. The pursuit of science shows

that there is a barrier, which no energy, mental or physical, can ever enable us to pass: that however profoundly we may penetrate the depths of space, there still remain innumerable systems, compared with which those which seem so mighty to us must dwindle into insignificance, or even become invisible; and that not only man, but the globe he inhabits, nay the whole system of which it forms so small a part, might be annihilated, and its extinction be unperceived in the immensity of creation. (vii)

This idea provides a perspective in which human existence and concerns are dwarfed by cosmic ones. Consequently, the pursuit of science is at once ennobling, empowering, and humbling.

Sixth, mathematics is the unifying language of the physical sciences, and the higher branches of mathematical analysis provide the medium in which "the extreme beauty" of the results of mechanical science is most easily visible. Despite its difficulties, it does not present an insuperable barrier; those who lack skill in the higher branches of analysis are not doomed to be ignorant of the results of such analysis or the methods by which the results are obtained. It is possible for them "to follow the general outline, to see the mutual dependence of the different parts of the system, and to comprehend by what means some of the most extraordinary conclusions have been arrived at." (vii) It is important to make "a sufficient distinction between the degree of mathematical acquirement necessary for making discoveries, and that which is requisite for understanding what others have done." (vii)

Seventh, science reveals the system of the world to be stable, enduring, and extremely well designed, despite the appearances of change and irregularity. The pattern of the universe is one of vicissitudes within immutability, chaos governed by order, irregularity within regularity. The system is marked by adaptation of means to ends, by stability, and by serenity.

Eighth, science is a heroic and progressive social enterprise, characterized by an ascent to knowledge through induction and an application, expansion, or transfer of knowledge through deduction. In this view, astronomy has advanced through the combined efforts of many astronomers in the face of adverse conditions and with "astonishing perseverance" from the earliest ages of civilization to the present. A series of steps led to "the knowledge of the universal gravitation of matter." Then, "descending from the principle of gravitation, every motion in the system of the world has been so completely explained, that no astronomical phenomenon can now be transmitted to posterity of which the laws have not been determined." (vi) This breadth of descriptive and predictive capability is the most impressive aspect of the "exact calculation" dimension of science.

Descriptive Patterns: The Cosmic Platform and Tracing the Mazes

The descriptive patterns Somerville uses to present the framework in the "Preliminary Dissertation" reinforce the themes and provide methods for developing explanations of scientific principles. There are two dominant ones: the cosmic platform and tracing the mazes. The cosmic platform is the point of view from which Somerville and her readers perceive the magnitude and splendor of the creation. From the cosmic platform, the heavens become a kind of Miltonic canvas against which the drama of the universe is played out.[4] The drama consists largely of tracing the mazes through which the operations of nature can be followed. The following passage from the beginning of the "Preliminary Dissertation" combines these descriptive patterns and sets the scene for all that will follow.

> The heavens afford the most sublime subject of study which can be derived from science: the magnitude and splendour of the objects, the

---

[4] See in *Paradise Lost*, for example, "Now had the Almighty Father from above, / From the pure Empyrean where he sits / High Thron'd above all highth, bent down his eye, / His own works and their works at once to view: / About him all the Sanctities of Heaven / Stood thick as Starrs, and from his sight receiv'd / Beatitude past utterance." (Book III, 56–62)

inconceivable rapidity with which they move, and the enormous distances between them, impress the mind with some notion of the energy that maintains them in the motions with a durability to which we can see no limits. Equally conspicuous is the goodness of the great First Cause in having endowed man with faculties by which he can not only appreciate the magnificence of his works, but trace, with precision, the operation of his laws, use the globe he inhabits as base wherewith to measure the magnitude and distance of the sun and planets, and make the diameter of the earth's orbit the first step of a scale by which he may ascend to the starry firmament. (Mech vi–vii)

Once the human observer has ascended through imagination "to the starry firmament," he or she enjoys a spectacular view: "the higher we ascend, the sky assumes a deeper hue, but in the expanse of space the sun and stars must appear like brilliant specks in profound blackness." (Mech lvi)

Astronomy is the most sublime subject in science, because of the power, immensity, and durability associated with the heavenly bodies and their movements and because it allows human beings to perceive and experience the presence of God. The greatness and goodness of the Creator are manifest not only in the sublime subject of the heavens but also in the human faculties that allow human beings to understand their intricate workings. This is in essence the credo of the scientific poets. Perceiving this intricacy is both exhilarating and humbling.

The cosmic platform is also a position from which human beings can look back at the earth. Following a discussion of the oscillations of the earth and moon that make it possible to see more than the hemisphere that is at any one time turned toward earth, she offers a view of the earth from the moon, as it might be seen by a "lunar traveller." From this view, the earth would appear as "a moon exhibiting a surface thirteen times larger than ours, with all the varieties of clouds, land, and water coming successively into view . . . a splendid object." (Mech xxxv)

The second dominant descriptive pattern, tracing the mazes, comes from Milton's *Paradise Lost* (Book V, 620–624) which Somerville quotes briefly in the "Preliminary Dissertation." (Mech xv)

> Yonder starry sphere
> Of planets, and of fix'd, in all her wheels
> Resembles nearest, mazes intricate,
> Eccentric, intervolved, yet regular
> Then most, when most irregular they seem.

Herschel echoes the same image with additional allusions to *Paradise Lost* in his review of *Mechanism* when he describes the mazy and mystic dance of the planets and their satellites – "cycle on epicycle, orb on orb" – from the earliest ages of which we have any record, nay, beyond all limits of human tradition, even to the remotest period to which all speculations carry us forward into futurity. (Herschel 1832: 537)

Herschel's description picks up both Milton's image of cycles within cycles and his reference to the "Mystical dance" of the planets.[5] These allusions to Milton are well suited to discussion of astronomy, which played an important role in Milton's thought, and locate the operations of the heavens in a distinctly theistic context.

Herschel demonstrates the way in which aesthetic principle and theistic ideas come together in his description of the motion of the orbits of the planets, which makes it clear that tracing the mazes is indeed following the movements of God.

The actual forms of their orbits are not ellipses, but spirals of excessive intricacy . . . yet this intricacy has its laws, which distinguish it from confusion; and its limits, which preserve it from anarchy. It is in this conservation of the principle of order in the midst of perplexity . . . that we trace the master-work-man, with whom the darkness is even as the light. (Herschel 1832: 541)

As this passage reveals, the theme of complexity and apparent chaos captured within a larger patter of simplicity and order draws much of its appeal from the irony or contradiction that it embraces. It was certainly not limited to Somerville's works but was used very effectively by her.

This pattern is used at the outset of the "Preliminary Dissertation" when Somerville speaks of tracing the magnificence of God's works with precision; she mentions it at the end of the "Preliminary Dissertation" as well. In both cases, it is clear that science is the primary means of tracing the mazes and thus an excellent way to perceive God.

The traces of extreme antiquity perpetually occurring to the geologist, give that information of the origin of things which we in vain look for in the other parts of the universe. They date the beginning of time [and] show that creation is the work of Him with whom "a thousand years are as one day, and one day as a thousand years." (Mech lxix–lxx)

In this case, as in many others throughout Somerville's works, tracing the mazes leads back to the images of the Psalms and gravity becomes emblematic of the Creator.

---

[5] The passage from *Paradise Lost* that Somerville quotes begins, "Mystical dance, which yonder starrie Spheare . . ." Quoted here from *The Poetical Works of John Milton* (London: Oxford University Press, 1925), p. 600.

Within the "Preliminary Dissertation," these themes and descriptive patterns are woven into a sophisticated account of the universe as revealed through physical astronomy. The bulk of the "Preliminary Dissertation" is devoted to technical description of which the examples so far quoted give little flavor. Since the technical descriptions are expanded to a much greater extent in *Connexion* and the themes and descriptive patterns are the same, this weaving process will be demonstrated much more extensively in the discussion of *Connexion*.

On the Connexion of the Physical Sciences

Somerville became intellectually, spiritually, and aesthetically engaged with tracing the forces of nature, especially gravity, through the universe, as she put together the view she presented in the "Preliminary Dissertation." Through her mastery of higher mathematics, she had perceived "the extreme beauty" of the results of analysis and seen the many ways that analysis and tools of observation, especially the telescope, could bring within the grasp of human beings a view of the universe that would otherwise have been beyond our reach. She had enjoyed success and gained confidence as an author and seen that she could be more than a transmitter of the work of others. In short, she realized that there was an audience for the intellectual perspective she had developed in the process of contextualizing Laplace. In *On the Connexion of the Physical Sciences*, she sought to trace the mazes farther. The world system portrayed in *Connexion* is a dynamic one, permeated by forces such as gravity that keep it constantly in motion. It is also permeated with a deep sense of mystery and the possibility that striking new revelations may come at any moment.

In its effect, *Connexion* was essentially the "Preliminary Dissertation" expanded to include a greater variety of phenomena and a broader range of new discoveries. In its content and function, *Connexion* was an extended state-of-the-art account of what was known at that time in physical science. Whewell characterized *Connexion* as "a *masterly* . . . exposition of the present state of the leading branches of the physical sciences." (Whewell 1834: 54) In his correspondence with Whewell, Herschel had spoken of the value of "digests of what is actually known in each particular branch of science" that give "a connected view of what has been done, and what remains to be accomplished in every branch." (Herschel quoted in Cannon 1978: 194) He believed that such works were important for directing future progress in science and for giving researchers a context in which to evaluate the originality of their own work. (Cannon 1978: 194) *Connexion* provided just such a digest and might be viewed as establishing a new literary form – the extended synthetic literature review.

In *Personal Recollections*, Somerville says, "In it [the "Preliminary Dissertation"] I saw such mutual dependence and connection in many branches of science, that I thought the subject might be carried to greater extent." (PR 178) In the "Preliminary Dissertation," these connections were left implicit; one of her tasks in *Connexion* was to make them explicit. This required tracing the phenomena of some parts of the natural world more deeply than she had before. It also required a great deal of additional research. She says, "There were many subjects with which I was only partially acquainted, and others of which I had no previous knowledge, but which required to be carefully investigated, so I had to consult a variety of authors, British and foreign." (PR 178–179) The challenges involved in assimilating and synthesizing the material conveyed in *Connexion* are reflected in Somerville's remark in *Personal Recollections* that "No one has attempted to copy my 'Connexion of the Physical Sciences,' the subjects are too difficult." (PR 294)

Connexion was both a single book and a series of books. The large number of editions produced reflected the need to keep the material up to date. *Connexion* not only served to disseminate scientific information; it also reflected and helped form consensus. Somerville worked with a very distinguished group of scientific consultants, whom Elizabeth Patterson characterizes as "possibly the most distinguished corps that any author ever commanded during a lifetime." (Patterson 1969: 331) The most important of her consultants included Herschel, Faraday, Wollaston, and Whewell. These and other scientists conveyed new findings, identified and explained important problems in their fields, and provided Somerville with their perceptions of the most current thinking in the various branches of science. Her approach to preparing successive editions resembled the refereeing process used by many scholarly journals today, and it appears that she both reported consensus and helped shape it through her writing.

Connexion provides convincing support for Somerville's claim that it was possible to follow the course of scientific discovery without extensive mathematical background. Although she presents a view of the physical world that was generated through highly mathematicized science, she uses no mathematical formulae or symbols in *Connexion*. This translation from mathematics into ordinary language was one of the most difficult aspects of the job. "Even the astronomical part was difficult," she says, "for I had to translate analytical formulæ into intelligible language, and to draw diagrams illustrative thereof, and this occupied the first seven sections of the book." (PR 179) Although this was the most difficult task she faced, it was equally the aspect that gave the book its greatest value over the long run.

She began with and gave the bulk of her attention to physical astronomy because it "affords the most extensive example of the connection of the physical sciences" (Conn 1) and also probably because it was the subject with which she was most familiar. She followed the general scheme of organization she had used in the "Preliminary Dissertation," adding detail to her discussion of many topics and including more extensive discussions of living systems and of electricity and related phenomena.

One of the most frequently asked questions about *Connexion* has been, What, exactly, does it connect?[6] This is an understandable question, given the rather unsystematic character of Somerville's writing. Her notebooks indicate that she sent her manuscripts to the printer in sections, and there is no evidence in her notebooks of outlines or other organizing devices for her writing. As in her previous work, her method of development in *Connexion* resembles the shifting of a gaze, as she moves from one subject of interest or point of focus to another. Although her writing has good local coherence, in that the individual portraits of phenomena and principles emerge clearly, there is no overall narrative structure or large-scale unfolding of logic. This is one of the most important senses in which her use of words is much more that of the poet or landscape artist than that of the storyteller or cartographer. At times she seems to be presenting a series of portraits of natural phenomena, and she uses phrases such as "next claims our attention" and "it is impossible to pass in silence" as she moves from one topic to another. Thus, she seems to be providing a visual survey or guided tour. As in the "Preliminary Dissertation," the individual scenes and passages of *Connexion* generally hold together well, but there is little overall continuity.

There is, however, overall unity. One of the primary ways the book achieves unity is through the interpretive frame she established in the "Preliminary Dissertation" and continues to use in *Connexion*. All the same elements are in *Connexion*, though they are slightly reordered. She still uses the descriptive patterns of the cosmic platform and tracing the mazes and adds a number of other explanatory techniques. The result of the changes is increased rhetorical force and a stronger emphasis on connection.

---

[6] James Clerk Maxwell addressed these questions in a review of *Grove's Correlation of Physical Forces* published in *The Scientific Papers of James Clerk Maxwell*, vol. 2 (Cambridge: Cambridge University Press, 1890), p. 401.

In a brief preface, she remarks, "The progress of modern science, especially within the last few years, has been remarkable for a tendency to simplify the laws of nature, and to unite detached branches of general principles." (n.p.) Her purpose is to demonstrate the range of these connections and how they have been discovered. She begins with astronomy because of the many kinds of connections it exemplifies. The following passage allows us to see the different kinds of connection, including the ways in which the intellectual pleasures of science can be linked to its aesthetic pleasures.

> Astronomy affords the most extensive example of the connection of the physical sciences. In it are combined the sciences of number and quantity, of rest and motion. In it we perceive the operation of a force which is mixed up with everything that exists in the heavens or on earth; which pervades every atom, rules the motions of animate and inanimate beings, and is as sensible in the descent of a rain-drop as in the falls of Niagara; in the weight of the air, as in the periods of the moon. (1)

In this passage, gravity continues to play an important role as a unifying force, but there are other forms of connection as well. As Somerville develops this idea, the reader traverses space and moves from "the sciences of number and quantity" to the perception of the universe as a giant, harmonious musical instrument.

Another example of her use of the pattern of tracing the mazes occurs in her description of the process by which light moves through the atmosphere. In the passage below, she is still tracing mazes from earth to the cosmic platform. Along the way, the reader is presented with a comparative analysis of the percentage of light rays reaching our eyes, which leads to an explanation of why we perceive things as we do. The reader thus ascends to the cosmic platform by means of a scientific understanding of the processes at work.

> The stratum of air in the horizon is so much thicker and more dense than the stratum in the vertical, that the sun's light is diminished 1300 times in passing through it, which enables us to look at him when setting without being dazzled. . . . Diminished splendor, and the false estimate we make of distance from the number of intervening objects, lead us to suppose the sun and moon to be much larger when in the horizon than at any other altitude, though their apparent diameters are then somewhat less. Instead of the sudden transitions of light and darkness, the reflective power of the air adorns nature with the rosy and golden hues of the Aurora and twilight. . . . The atmosphere scatters the sun's rays, and gives all the beautiful tints and cheerfulness of day. It transmits the blue light in greatest abundance; the higher we ascend, the sky assumes a deeper hue; but in the expanse of space, the sun and stars must appear like brilliant specks in profound blackness. (Conn 152–153)

She concludes with a statement that exemplifies the ways the aesthetic pleasures of science are joined to the intellectual ones:

> It is impossible to thus trace the path of a sunbeam through our atmosphere without feeling a desire to know its nature, by what power it traverses the immensity of space, and the various modifications it undergoes at the surfaces and in the interior of terrestrial substances. (Conn 153)[7]

One of the strengths of Somerville's writing is that it allows the reader to see science at work in the world, and she often begins an explanation of a complex principle by presenting a familiar example in which the principle is at work, as in the following explanation of the tides. This passage also illustrates another technique she employs, creating alternative scenarios – usually simpler than what actually occurs – and then using the simpler case as a context for introducing a complication that occurs in reality.

> One of the most immediate and remarkable effects of a gravitating force external to the earth, is the alternate rise and fall of the surface of the sea twice in the course of a lunar day. . . . If the moon attracted the center of gravity of the earth and all its particles with equal and parallel forces, the whole system of the earth and the waters that cover it would yield to these forces with a common motion, and the equilibrium of the seas would remain undisturbed. (85–86)

But, she tells us, this simple case is not the reality: "On the contrary, the moon attracts the center of the earth, more powerfully than she attracts the particles of water in the hemisphere opposite to her; so that the earth has a tendency to leave the waters, but is retained by gravitation, which is again diminished by this tendency." (Conn 87)

We see another example of this "if the simpler case obtained" explanatory strategy in the following passage that deals with the motions of the planets:

> Were the planets attracted by the sun only, they would always move in ellipses, invariable in form and position. . . . The true motions of the planets are extremely complicated, in consequence of their mutual attraction; so that they do not move in any known or symmetrical curve, but in paths now approaching to, now receding from, the elliptical form. (Conn 10–11)

---

[7] It is interesting to consider these passages in light of Jones' argument that the scientific sublime collapsed under the didactic burden (1966: 200–212); they are clearly examples of scientific prose that easily carries the didactic burden at the same time that it conveys the scientific sublime.

She also uses the technique of presenting alternative scenarios when she discusses the velocity at which the celestial bodies "were first propelled in space."

> Had that velocity been such as to make the planets move in orbits of unstable equilibrium, their mutual attractions might have changed them into parabolas, or even hyperbolas; so that the earth and planets might, ages ago, have been sweeping far from our sun through the abyss of space. But as the orbits differ very little from circles, the momentum of the planet, when projected, must have been exactly sufficient to insure the permanency and stability of the system. . . . There is not in the physical world a more splendid example of the adaptation of means to the accomplishment of an end, than is exhibited in the nice adjustment of these forces, at once the cause of the variety and of the order of Nature. (Conn 12)

This last example illustrates another characteristic of the explanatory pattern of tracing the mazes; it almost always leads to God.

One of her most useful techniques for clarification is one whose value has only recently come to be recognized by scholars in technical communication – recognizing and then correcting reasonable but incorrect inferences.

> It might be imagined that the reciprocal action of such planets as have satellites would be different from the influence of those that have none. But the distances of the satellites from their primaries are incomparably less than the distances of the planets from the sun, and from one another; so that the system of a planet and its satellites moves nearly as if all these bodies were united in their common center of gravity. (Conn 25)

This technique is especially important for maintaining the accuracy of the view presented to the imagination. Imagination is a powerful force to engage but not all that easy to direct. The technique of recognizing and correcting reasonable but incorrect inferences is particularly useful in this regard.

This technique is related to one of the most interesting aspects of her philosophy of science, her belief in "the fallacy of our senses," to which science is an important corrective. This might at first seem contradictory, since science is founded on observation, information gained either directly or indirectly through the senses in most cases. But the following passage indicates how science can exercise this corrective power, even within a philosophy that insists that all knowledge of the external world is founded on experience.

> A consciousness of the fallacy of our senses is one of the most important consequences of the study of nature. This study teaches us that no object is seen by us in its true place, owing to aberration; that the colors of sub-

stances are solely the effects of the action of matter upon light; and that light itself, as well as heat and sound, are not real beings, but mere modes of action communicated to our perception by the nerves. (Conn 230–231)

This emphasis on the fallacy of the senses helps reinforce one of the most important aspects of the belief system in which Somerville operates – the idea that some mysteries still remain and will likely always be there. This is important because, in the tradition in which Somerville is working, mystery and the unknown are associated with God. A science that emphasizes mystery and the unknown is thus much more compatible with a theistic view.

Another interesting feature of Somerville's philosophy of science is the idea that apparently trifling phenomena lead to major conclusions or to the perception of analogies and connections of importance. This idea is developed in the following discussion of polarization:

the original discovery of that property of light was accidental, and . . . would have passed unnoticed, had it not happened to one of those rare minds capable of drawing the most important inferences from circumstances apparently trifling. In 1808, while M. Malus was accidentally viewing with a doubly-refracting prism a brilliant sunset reflected from the windows of the Luxembourg palace in Paris . . . he was surprised to see a very great difference in the intensity of the two images. . . . A phenomenon so unlooked for induced him to investigate its cause, whence sprung one of the most elegant and refined branches of physical science. (Conn 189)

This is an example of the kind of discovery that led Somerville and her contemporaries to believe in the value of collecting large amounts of detailed observations, even if their immediate significance was not clear.

One of the most powerful forms of reasoning available to the scientist was the perception of an analogy between phenomena. This was the means by which Newton had gained his dramatic insight regarding gravitation and one of the most powerful methods of advancing science as well as showing evidence of individual creativity. Examples of analogies were important because a knowledge of analogies perceived by others could help attune one's mind to the perception of analogies not yet recognized. Somerville was extraordinarily good at drawing parallels and making analogies that exploited similarities without losing sight of the distinctness of the entities being compared.[8]

---

[8] Another example of this occurs in *Connexion*, pp. 142–143, in Somerville's discussion of recurring vibrations.

Somerville's ideas about the interdependence of natural phenomena are illustrated most extravagantly in the closing passages of *Connexion*, where she weaves together all the connections established within the text in what is a sometimes dizzying account of tracing the mazes.[9] Following these elaborate mazes reveals another important principle of her philosophy of science – that we can learn a great deal from small amounts of apparently unrelated information once we have perceived that a connection exists, and that knowledge of one phenomenon usually leads to knowledge of other phenomena. For example, a "knowledge of the action of matter upon light" leads to an understanding of the true locations of both earthly and heavenly objects and the nature and properties of the sunbeam. Similarly,

> the effects of the invisible rays of light are immediately connected with chemical action; and heat, forming a part of the solar ray so essential to animated and inanimated existence . . . is too important an agent in the economy of creation, not to hold a principal place in the connection of physical sciences. Whence follows its distribution in the interior and over the surface of the globe, its power on the geological convulsions of our planet, its influence on the atmosphere and on climate and its effect on vegetable and animal life. . . . The connection of heat with electrical phenomena, and the electricity of the atmosphere, together with all its energetic effects, its identity with magnetism and the phenomena of terrestrial polarity, can only be understood from the theories of these invisible agents, and are, probably, identical with, or at least the principal causes of, chemical affinities. Innumerable instances might be given in illustration of the immediate connection of the physical sciences, most of which are united still more closely by the common bond of analysis, which is daily extending its empire, and will ultimately embrace almost every subject in nature in its formulæ.
>
> These formulæ, emblematic of Omniscience, condense into a few symbols the immutable laws of the universe. This mighty instrument of human power itself originates in the primitive constitution of the human mind, and rests upon a few fundamental axioms, which have eternally existed in Him who implanted them in the breast of man when He created him after His own image. (Conn 389–390)

As in the other instances, tracing the mazes leads back to the Creator and to the powers of perception that man shares with the Creator. The view that emerges from *Connexion* is one in which all things are linked through the physical universe and the divine mind.

To grasp Somerville's intellectual project, it is essential to understand that the connections that she first perceived in writing the "Pre-

---

[9] The full passage appears in "The Nature of Her Story" in chapter six.

liminary Dissertation" and that she explored in *On the Connexion of the Physical Sciences* preexisted science. They were inherent in the entity conceptualized by Somerville and many of her contemporaries as the universe or cosmos, a complete, harmonious, and orderly system. Within that conception – in their idea of the universe – the connections of the various aspects of animate and nonanimate nature did not have to be *established* by science; they had only to be discovered and traced. Because of the vastness of the universe and the complexity of interactions, this task was an enormous one, as Somerville expressed it in a letter of August 29, 1841, to her son Woronzow, an undertaking "more fit for the combination of a Society than for a single hand to accomplish." (PR 249) Unless a single individual undertook it, however, it was unlikely that the synthesis required for unity could be accomplished.

When James Clerk Maxwell reflected on Somerville's period later in the century, he characterized it as one in which "there already existed a widespread desire to be able to form some notion of physical science as a whole." (Maxwell 1890: 401) Although Maxwell clearly respected Somerville and admired her writing, he seems disappointed at the kinds of connections she established, concluding, "The unity shadowed forth in Mrs. Somerville's book is therefore a unity of the method of science, not a unity of the processes of nature." (402) He seems to be looking for another kind of connection than those Somerville elucidated and that have been discussed here.

But Maxwell himself recognized the value of works that synthesized and focused the thinking of the scientific community. "It is not by discoveries only, and the registration of them by learned societies, that science is advanced. The true seat of science is not in the volume of Transactions, but in the living mind, and the advancement of science consists in the direction of men's minds into a scientific channel." (402) Maxwell puts works that perform this function into the category of "suggestive books," whose function is to "put into a definite, intelligible, and communicable form, the guiding ideas that are already working in the minds of men of science, so as to lead them to discoveries, but which they cannot yet shape into a definite statement." (402) We might think of these guiding but not yet definitely stated ideas as emerging consensus or theories. We might also think of the ways in which Somerville's observations suggested lines of inquiry to other scientists. It appears that one of Somerville's observations in *Connexion* led one of the codiscoverers of Neptune, John Couch Adams, to begin the investigation that led to his discovery of the planet. Somerville had suggested that the perturbation of Uranus might indicate the existence of a planet that had not yet been observed, and Adams' work was in part motivated by that suggestion. (Grosser 1979: 79)

As *Connexion* evolved through many editions, it reflected evolving scientific consensus. Somerville has sometimes been criticized for being too "flexible" in changing her views over time (Grosser 1979: 52–54), but such a criticism misses the point of what she was doing: her views changed because science progressed. When new theories were debated, she reflected the debates; when new information came to light, she speculated on what it might mean. The various editions of *Connexion* thus provide a historical review of the various forms "into which science was shaping itself." (Maxwell 1890: 402) Her works played a significant role in the forming of scientific opinion, which Maxwell described as "the influence which determines what a scientific man shall say when he has to make a statement about a science which he does not understand." (402)

Maxwell gives Somerville no credit for contributing to the development of the idea of the conservation of energy, but, when Thomas Kuhn (1962) reassessed the situation many years later, he saw things differently. Kuhn relates Somerville's ideas as expressed in *Connexion* to a number of developments that occurred after 1830, when scientists began to perceive connections among phenomena previously thought to be unrelated. Referring to Somerville's recognition of the connection between the sciences, that is, her belief that "there exists such a bond of union, that proficiency cannot be attained in any one branch without knowledge of others" (Conn preface), Kuhn says,

Mrs. Sommerville's [sic] remark isolates the "new look" that science had acquired between 1800 and 1835. That new look, together with the discoveries that produced it, proved to be a major requisite for the emergence of energy conservation. Yet, precisely because it produced a "look" rather than a single clearly defined laboratory phenomenon, the availability of conversion processes enters the development of energy conservation in a variety of ways. . . . In short, just because the new nineteenth-century discoveries formed a network of "connexions" between previously distinct parts of science, they could be grasped either individually or whole in a large variety of ways and still lead to the same ultimate result. . . . What Mrs. Sommerville [sic] had called the new "connexions" between the sciences often proved to be the links that joined disparate approaches and enunciations into a single discovery. (1962: 324–326)

This argument suggests that the process of discerning links or "connexions" is an intermediate stage in which discrete discoveries are associated in a way that eventually leads to large-scale synthesis. This intermediate stage is perhaps less recognized than the other two ends of the scale but plays an essential role in the overall advancement of science.

These Works in the Context of Somerville's Career

With these details of *Mechanism*, the "Preliminary Dissertation," and *Connexion* in mind, it is fruitful to consider what these works mean in the context of Somerville's career as a scientist and what they illustrate about the peculiar illumination of the female mind, the role of the feminine in her writing.

Her first three major publications and the response they received provide evidence of the eminence she had achieved and the innovation of which she was capable. The response of her colleagues also reflects a belief in her capacity for original contributions in science. In his review of *On the Connexion of the Physical Sciences*, David Brewster publicly urged her to extend her own original investigations: "Mrs. Somerville's great mathematical acquirements, her correct and profound knowledge of the principles of physical science, and the talent for original inquiry which she has already evinced in her paper on the magnetism of the solar rays, induce us to urge her to original investigation." (1834: 171) The belief in Somerville's ability to make original contributions was essential to the high place she held within the scientific community. Although there might be debate about the long-term value of her original contributions, she had clearly demonstrated the capacity for original work.

At the end of the same review, Brewster offered a view of authorship that seems somehow at odds with the praise he has heaped on Somerville but that offers insight into the difficulty inherent in the place she had created for herself: "The fame of scientific authorship is but a poor compensation for its toils; and the fleeting celebrity of writing the best book upon a science which is undergoing continual change, and demanding new expositors, cannot gratify a mind like hers." (1834: 171)

Remarks Somerville made later in life suggest that he may have been right to a limited extent about "scientific authorship" failing to gratify a mind like hers, but he was undoubtedly correct in his conclusions about its "fleeting celebrity." Even in Somerville's day, the value of the kind of writing she did depended on its currency. Although state-of-the-art information was conveyed in books in Somerville's day (as opposed to the periodical literature that carries such information today), frequent editions had to be produced. Even the best writing quickly became out of date (as the fate of even the great Laplace demonstrates). Though Somerville had a remarkably long career as a successful writer, "the fame of scientific authorship" proved fleeting indeed, especially as the kind of synthetic intellectual enterprise in which she was engaged began to give way later in the century to the disciplinary approaches of newly professionalized, highly specialized scientists. The space she occupied offered an avenue to eminence, but it was not a permanent opening, or a role that could easily be assumed by other women.

Her first three major scientific works also reflect the ways in which Somerville had finally been able to make space within her domestic life for concentrated and sustained, though still interrupted, scientific work. As a middle-class woman, she had always enjoyed a fair amount of time for intellectual pursuits. As she moved past her child-bearing years, her children became more independent. Her oldest child, Woronzow Greig, had taken his degree from Cambridge in 1826, and her daughters were gaining a measure of independence, though they still took up a great deal of her time. She made the most of that space by virtue of good personal discipline, extraordinary powers of concentration, and lifelong habits of maintaining a diversity of pursuits.

An additional factor that helped create space and maintain momentum was the need to make money, which played a somewhat complex role in Somerville's career. The Somervilles had a number of financial crises that hit their peak in 1835 as a result of debts incurred by a cousin and guaranteed by William Somerville without Mary Somerville's knowledge. Over the years, although the sense of crisis abated, the need to make money remained.[10] Like other women authors, she gained a certain freedom from the necessity of bringing in income at the same time that she felt pressured to present her writing as something other than income-generating activity.[11] The culture of science at the time she established herself allowed her to occupy a comfortably ambiguous space in which she could publicly and plausibly protest that she wrote only for pleasure while privately paying considerable attention to the profitability of her books. Moreover, pleasure, self-satisfaction, and the need to make money seem to have played equally important roles in her career.

---

[10] For a detailed discussion of the financial situation and problems of the Somerville family, see Patterson 1983, especially pp. 170–172.

[11] See David 1987. Harriet Martineau is an interesting figure to compare with Somerville. Like Somerville, Martineau welcomed the opportunity to do something useful and to be paid for it. Martineau apparently felt more subservient in her role than Somerville did in hers.

The available evidence suggests that she probably did not see her first publishing venture with *Mechanism* as an opportunity to make money. *Mechanism* was a critical success, but it sold too few copies to be a financial success. *Mechanism* did, however, allow her to establish herself as a credible scientist with a clear audience for her work. Through it she developed a relationship with a publisher (John Murray) and subsequently was in a position to envision what might sell. When she had begun *Mechanism* in 1827, she had, as Patterson expresses it, already acquired "some of the aura of a scientific elder." (1983: 68) Younger male scientists looked at her as a valuable consultant. Brougham's invitation both reflected the esteem in which she was held among the growing group of scientists who knew her personally and offered the opportunity for greater public exposure. Although going public as an author put her at risk, she had not only lived up to her previous reputation but had also greatly enlarged it and put herself in a position to earn a significant portion of the family living. In 1835 following the publication of *Connexion*, she was put on the Civil List with £200 per year, which was increased to £300 in 1837. This amount is comparable to that given to Airy, Faraday, Dalton, Brewster, and Ivory and reflected her standing within the scientific community.[12] Another advantage of her pension and her career as an author was that they provided her with an ongoing incentive to maintain and expand her range of scientific contacts. These factors were more important than they might at first seem given the pressures that deterred women from intellectual careers during Somerville's day.

Her first three major publications also reveal something important about the many roles she played in relation to her readers. The Mary Somerville who emerges in these works is a scientific expert, a rhetorician, and a philosopher with a great talent for vivid descriptive writing. She allows the reader to look at science from many different perspectives and to witness phenomena on a scale and with an intensity that expands the readers' conceptions. She allows her readers to see through the eyes of Laplace and dozens of other scientists and thus greatly enriches their views of the world. In relation to the scientific community, she both articulates consensus and identifies areas of controversy, which she enters into on occasion although always in a measured and careful way. She revises opinions as new facts emerge. Viewing successive editions of her works gives one a sense of the fluidity of knowledge, the way it acquires validity and evolves over time. She emerges

---

[12] There was some political controversy relating to the pensions of Somerville and other cultivators of science. For a full discussion of this subject, see Patterson 1983, especially chapter 8, "The Civil List and Mary Somerville," pp. 151–162.

as a philosopher in the sense that she articulates the overarching principles and assumptions that underlie the scientific enterprise. This philosophical bent is somewhat difficult to discern because she is not very systematic in developing it. She does, however, articulate most of the major truths revealed by science as well as the major principles on which science operates. To accomplish these varied objectives, she assumes different relationships to her readers and her subject, always keeping herself appropriately in the background.

The responses to her work do not indicate that she was allowed any kind of license by virtue of being a woman. In fact, the situation seems to be quite the opposite. As Herschel's review of *Mechanism* demonstrates, her writing was not judged by the standards for "women's writing" but rather by the standards for "scientific writing." But as Herschel's comments also illustrate, she not only had to exhibit the positive characteristics of scientific writing but also avoid the errors presumed to be common to female authors. Thus, gendered conceptions very much entered into the evaluation of her work. The reviewers looked for negative traits relating to her gender, and, finding none, proceeded to judge her by the standards that would have been applied to any scientist. This is an essential point for understanding what she achieved.

The question remains whether there is anything distinctively feminine about her writing. There does seem to be a sense in which the scientific sublime may have been particularly appealing to women, as the comments of Edgeworth and other women suggest. In his discussion of the Romantic sublime, Weiskel emphasizes the liberating nature of the sublime experience as articulated by W. Jackson Bate: "Through the personal rediscovery of the great, we find that we need not be the passive victims of what we deterministically call 'circumstances' (social, cultural, or reductively psychological-personal), but that by linking ourselves . . . with the great we can become freer." (In Weiskel 1986: 11)

The association of the sublime with becoming freer and transcending limitations may have conveyed a welcome sense of freedom for Edgeworth, Somerville, and other women, especially given the relative lack of freedom that characterized their lives. And it is possible that she presented the scientific sublime in a way that made it more accessible to women readers than it would have been otherwise. Confined as they were by the dictates of their culture, these women may have been particularly open to the appeal of the scientific sublime because of the sense of freedom and empowerment they derived from it. On the other hand, there is ample evidence to suggest that men found the scientific sublime very appealing as well.

In the end, there does not seem to be anything stereotypically feminine about Somerville's work. In her scientific writing, nature is active, and she watches it intently and portrays it dramatically. She is neither a passive observer, nor an aggressive one, but instead interactive, a participant. She enters into nature imaginatively and analytically and gains a firm conceptual grasp of most of the phenomena she observes. The vividness of her portrayals derives in large part from her ability to weave together the discourses of drama, science, poetry, aesthetics, philosophy, and theology. She weaves these together into a kind of three-dimensional tapestry that portrays an active world as it is viewed scientifically. This weaving gives her accounts, at least in places, the richness captured in Whewell's phrase, "an expanse magnificently occupied." The rhetorical and literary dimensions of her work are particularly important. By approaching science from so many different perspectives, she gives a picture of science that includes its methods, subject matter, major conclusions, and motivations; its practical and philosophical significance; and its aesthetic as well as its intellectual pleasures. Her accounts are rich, accurate, and comprehensive, accessible to non-experts and acceptable, indeed welcomed, by experts.

The multiplicity of discourses related to science has been increasingly recognized in recent years. Based on his study of scientific writing done in the field of evolutionary biology, Greg Myers (1990) has concluded that the distinction between even very sophisticated popularization and specialized writings is that the specialized writing focuses on the process of science, the disciplinary procedures involved, rather than the objects of study that are the focus of popularizations. Although Myers rejects the tendency to see either popularization or specialist discourse as a distortion of the other (141), he believes that there are two distinct kinds of narratives: a narrative of nature associated with popular writing and a narrative of science associated with specialized writing. (189–190)

The narrative of nature focuses on the subject of scientific study such as a plant, an animal, or, as would be the case in much of Somerville's writing, a natural force or phenomenon. The narrative of nature tends to be chronological and emphasizes the externality of nature to scientific processes. The narrative of science, on the other hand, focuses on the arguments of scientists and scientific activity and procedures, and tends to arrange time into simultaneous, parallel sets of events that follow the conceptual structure of the discipline.

The purpose of the narrative of science, in Myers's terms, is "to integrate researchers and their findings into the work of the research community" and to show that they meet the standards and accept the constraints imposed by the discipline. "Each article is a demonstration of the need for scientific expertise" in the sense that it emphasizes the need for making well-informed choices. (1990: 189) The narrative of nature, on the other hand, emphasizes simple observations, "minimizes expertise and emphasizes the unmediated encounter with nature. All scientific knowledge is brought within the realm of common sense, all scientific knowledge serves public goals." (1990: 191) Following Myers, Barbara T. Gates and Ann B. Shteir (1997) identify a third kind of narrative, the narrative of natural theology. As previously discussed in chapter one, this narrative presumes that "describing the wonders of nature" is at the same time "describing the wonders of creation and ultimately, of a creator." (11)

As the passages of Mary Somerville's writing quoted here and the discussion of her framing technique demonstrate, her writing blended all three of these narratives – nature, science, and natural theology – in a very interesting and sophisticated way. The narrative of natural theology is a significant part of the interpretive framework she creates. Within the frame of a narrative of natural theology, she blends the narrative of science with the narrative of nature. Much of her writing focuses on the forces in nature, which are often portrayed as quite active agents. The drama played out in her writing is the drama of those forces. In these aspects of her writing, she is working within what Myers calls the narrative of nature. She also uses the narrative of science and devotes time to the methods and processes of science, to controversies, debates, and emerging as well as accepted theories. Her reader is aware that knowledge must be produced and that there are choices of both theory and interpretation involved. She does not offer the reader an unmediated encounter with nature. It is an encounter that is heightened because it is mediated by science and by Somerville's vision of the aesthetic and spiritual pleasures of science. This blending and balancing is another example of the principle of counterpoise at work.

In the final analysis, it seems that Somerville's writing was "peculiarly" feminine because it was not clearly recognizable as feminine at all. It neither violated norms of femininity, nor exhibited the negative traits associated with women writers. Gendered ways of thinking were so important and the masculine identity of science and other higher intellectual pursuits already so strongly entrenched, that it would probably have been impossible for Somerville to have somehow transcended gendered thinking altogether. But it was possible for her to offer a third alternative. To borrow Rose's terminology, Somerville works not in Bacon's narrative "of the connection of knowledge and power" nor in the feminist narrative "of the danger of knowledge without love." Her narrative instead emphasizes interdependence and connection as fundamental principles of nature, of science, and of human experience, and, perhaps most importantly, the power of relation and connection. She thus blended gendered traits and blurred gendered categories in a way that allowed her to occupy a distinctive and prominent space – that of the eminent scientist.

4

The Earth, the Sea, the Air, and Their Inhabitants

Physical Geography *(1848) and* On Molecular and Microscopic Science *(1869)*

The deeper the research, the more does the inexpressible perfection of God's works appear, whether in the majesty of the heavens, or in the infinitesimal beings of the earth.
> – Mary Somerville, *On Molecular and Microscopic Science*

As in a theater,
The lights are extinguished, for the scene to be changed
> – T. S. Eliot, "East Coker"

As previous chapters have demonstrated, one of the hallmarks of Somerville's rhetoric of science was a theme she borrowed from the eighteenth-century scientific poets – the idea that science was a pathway to God, a form of elevated meditation, and that detailed examination of nature revealed the intricacy, drama, harmony, and beauty that God had incorporated into the design of universe. In this view, the capacity for exact calculation and for in-depth and precise understanding of phenomena was an aid rather than a hindrance to appreciating the wonders of the creation. She pursued the creative possibilities of this idea in new settings in *Physical Geography* and *On Molecular and Microscopic Science*. In *Mechanism* and *Connexion*, Somerville had provided her readers with an expanded conception of the universe and surveyed the celestial and some aspects of the terrestrial spheres. In *Physical Geography* and *On Molecular and Microscopic Science*, she turned her attention to the remainder of God's handiwork – the earth, the sea, the air, and their many animal and vegetable inhabitants. As she explored and presented detailed scientific accounts of these subjects, Somerville took up another theme borrowed from the scientific poets: the analogy between the worlds revealed by the telescope and microscope. She also took up the challenge of portraying the sublime as it is manifest in apparently mundane or invisible phenomena. Though the drama she unfolded for the reader was in many ways the same, the scene and the aesthetics were different.

In *Physical Geography* and *On Molecular and Microscopic Science*, the projects that engaged much of Mary Somerville's attention during the last thirty-four years of her life, she greatly expanded the range of natural phenomena treated. This expansion often took her beyond her own direct experience of science, but it also offered rich opportunities for presenting exciting new discoveries and for exploring the aesthetic dimensions of science. In these works, she offered extensive and detailed treatment of the terrestrial portion of creation, where the challenges of understanding the processes at work were less daunting but the challenges of making the material awe-inspiring and providing the reader with a sense of having *witnessed* the phenomena described were greater. She also reached out to an increasingly broad audience.

A Change of Scene

These changes in subject and approach may in some ways reflect the changes of scene and circumstance that Somerville herself experienced. As the narrative of chapter one indicated, the Somerville family had moved to Italy in 1838 in the hope of improving William Somerville's health. They settled first and most frequently in Rome but moved from place to place – Naples, Florence, Lake Como, Siena, Genoa, Bologna, Sorrento, Spezia. Florence offered the best scientific society. The physicists G. B. Amici and Vincente Antinori were there, and Tuscany's ruler, the Grand Duke Leopold II, who had been elected a Fellow of the Royal Society in 1838, offered her the use of his excellent library at the Pitti Palace. But the climate at Rome seemed best for William Somerville's health.

Mary Somerville was well received by Italian scientists but did not have as much interaction with the scientific community as she had enjoyed in Britain. She continued to keep up with British science through correspondence and visits but necessarily had less direct contact with British researchers. She had to rely more on published reports and books and less on correspondence and personal interaction with other scientists. She was also under increasing pressure to write books that would sell as William Somerville was retired at half-pay, and there were other drains on the family finances. These factors of relatively constant travel and the exposure to new scenery and cultures seem to have increased her awareness of the variety in the landscape, wildlife, and plant life in different geographical areas. This new awareness, combined with a need to rely on published as opposed to private sources of information, led to *Physical Geography*, the book that sold more copies than any of the others.

If we consider *Physical Geography* and *Molecular and Microscopic Science* as extending a project begun with *Mechanism* and *Connexion*, we can see Somerville working her way through the cosmos – from the world revealed through the telescope to that revealed through the microscope, from astronomy to botany, chemistry, physiology, and molecular physics. She provided a complete description of the physical universe in all its vastness, complexity, and variety. Although Somerville did not characterize this larger enterprise herself, it might be conceived as a cosmography – a study of the entire universe and its related parts, encompassing not only astronomy, geography, and geology but also the inhabitants of the earth, sea, and air, and their multiple interactions with the environment. She also fully explored the capacity of that universe to evoke the aesthetic responses that result when human beings perceive and appreciate its unity. In her view and the view of many of her contemporaries, that unity did not have to be created and did not derive from any structure she imposed. Rather, it preexisted science, and science simply served as a way of comprehending it more fully. Both scientific law and aesthetic pleasure resulted from recognition of that unity.

Physical Geography: *An Overview*

Physical geography, as Somerville defined it, included more than we ordinarily associate with geography. For example, she included geology and the distribution of animal and vegetable life. Her objective was to discern the most important connections and thus to determine the principles that shaped the land and explained the distribution of animal and vegetable life on land and in water. She also sought to understand the various transformation processes involved. Cannon refers to this as "Humboldtian science," a field that "includes astronomy and the physics of the earth and the biology of the earth all viewed from a geographical standpoint, with the goal of discovering quantitative mathematical connections and interrelationships." (Cannon 1978: 77; see also Dettelbach 1996) One significant aspect of this approach was that it also included human beings. Somerville indicated that she was "following the noble example of Baron Humboldt, the patriarch of physical geography" and taking "a more extended view of the subject than the earth and its animal and vegetable inhabitants." This broader view included "the past and present condition of man, the origin, manners, and languages of existing nations, and the monuments of those that have been." (PG 466) Because it included human and other living beings, *Physical Geography* took Somerville into new and interesting territory.

Although it has sometimes been said that *Physical Geography* offered Somerville her first opportunity to voice her own opinions, she had in fact voiced opinions from the beginning of her career as an author. In *Mechanism*, she had commented on the value of the work of Laplace and other important contributors to mathematical science. In *Connexion*, she offered numerous judgments about the validity of various scientific theories and findings. She had attempted to articulate the emerging or established consensus of the scientific community in her first two books, and continued to do this in *Physical Geography*. What is different is that she is discussing human and contemporary affairs in a fairly extended way. This broadening of subject reflects her growing sense of confidence as an author as well as the authority she carried with her contemporaries. In her previous works, she had referred to technological developments, explained their significance, and articulated a theisitic framework for science, but she had not speculated on the direction of human affairs to nearly the extent she did in *Physical Geography*. In this third book, she articulated many of the commonplace beliefs of her day and took some dissenting views as well.

Her comments reveal a striking similarity in the models she used for thinking about the natural and the social worlds.

In *Physical Geography*, as in her other works, Somerville offers a short statement of purpose, defines her subject briefly, establishes the interpretive framework she will use throughout the book, and then launches into presentation of the content. Relatively little space is devoted to the development of the framework. The first chapter provides a sketch of the subject but says almost nothing about her purpose in writing or what will follow. She offers few organizing statements, and the reader gets little sense of the completeness of the overall design of the work or of where she is in her coverage of the material. The logic behind the sequence in which topics are taken up is never made explicit. She indicates an interest in natural history early on, which is expressed as an interest in "the former state of our terrestrial habitation, the successive convulsions which have ultimately led to its present geographical arrangement, and to the actual distribution of land and water, so powerfully influential on the destinies of mankind." (PG 13) Still, there is no extended historical narrative. As before, she offers what might be characterized as local descriptions – descriptions of the conditions or distributions at a particular place and what led up to them.

After describing the overall structure of the earth, in the "Outline of Geology" that makes up most of chapter 1, she moves on to other features of the earth, beginning with the forces that raised the continents. She makes brief allusion to the location of the earth within the solar system but quickly focuses on terrestrial topics. She starts with the most basic features of the land and water and the major formations found in them, such as mountains, volcanoes, and mineral veins. She then treats the more dynamic elements of the natural landscape – oceans, tides, rivers, and lakes – and the forces at work in them, including those that govern temperature, light and color, electricity, storms, the aurora, and magnetism. In chapter 23, she turns to vegetation, which she describes according to its geographical distribution, and then to insects, fish, and reptiles, again organized according to geographical distribution. Chapter 31 is devoted to birds discussed by regions – Arctic, European, Asian, African, American, and Antarctic – and then mammals. She concludes the book with an extended discussion of "The Distribution, Condition, and Future Prospects of the Human Race." Human beings are an essential aspect of physical geography as she defines it because they are both profoundly influenced by as well as profound influences on the environment.

She supplements her descriptions with tables of data and footnotes that offer extensive discussion of controversial issues or issues only remotely related to the subject at hand. She often gives the sources of her information or directs the reader to additional information on a subject. The overall objective seems to be to provide as much precision as possible without breaking the reader's momentum with unnecessary detail. By and large, she succeeds in achieving this objective.

The Interpretive Frame in Physical Geography

Although the interpretive frame used in *Physical Geography* relied on the same basic belief system and philosophy of science underlying *Mechanism* and *Connection*, its themes and descriptive patterns were adapted to suit and to exploit the possibilities of the new subjects at hand. The science conveyed in *Physical Geography* is based more on observation than calculation but still progresses through analogy, comparison, and generalization. The power associated with the predictive capability of astronomy or the pervasiveness of the law of gravity finds a ready substitute in the forces that shape the earth and its inhabitants and in the activity with which Somerville could endow an apparently inert landscape.

The interpretive frame for *Physical Geography* briefly recalls the cosmic view of *Mechanism* and *Connexion*, reiterating that the earth's position in the solar system and its connection with other celestial bodies "have been noticed by the author elsewhere. It was there shown that our globe forms but an atom in the immensity of space, utterly invisible from the nearest fixed star, and scarcely a telescopic object to the remote planets of our system." (PG 13–14) But she spends very little time presenting the view from the cosmic platform. The ordinances of the heavens and the immensity of outer space are replaced by the story of the solid earth and the forces that seethe beneath it. The Miltonian vision remains, but this time it is the burning lake instead of the starry spheres, the language and images of the scientific poets overshadowed by those of Genesis and the Psalms. There are still images of motion, of a dynamic universe, but the scene is very different. In a passage reminiscent of the burning lake in *Paradise Lost*, Somerville describes "the increase of temperature with the depth below the surface of the earth, and the tremendous desolation hurled over wide regions by numerous fire-breathing mountains, [which] show that man is removed but a few miles from immense lakes or seas of liquid fire." (PG 14) She emphasizes the instability and fragility of the "shell" of the earth, the power and activity beneath, and works within a strain of the sublime that takes pleasure in the perception of destructive power. Although the play seems to be the same, the setting is a new one.

The Themes

The central theme in the interpretive frame of *Physical Geography* is a unitary conception of the world, a view based not on political or other "arbitrary divisions" created by human beings but rather on the way things have been structured by the Creator. Choosing the subject of physical geography as her vantage point, she provides a comprehensive view.

> Physical Geography is a description of the earth, the sea, and the air, with their inhabitants animal and vegetable, of the distribution of organized beings, and the causes of their distribution. Political and arbitrary divisions are disregarded, the sea and the land are considered only with respect to those great features that have been stamped upon them by the hand of the Almighty. (PG 13)

This theme transforms the notion of God as "Almighty Architect" and adapts it for a more terrestrial scale of consideration. Although *Physical Geography* contains occasional references to "Divine Wisdom" and the "Creator," it puts much more emphasis on "Providence," a concept that is related but differs in some significant ways. "Providence" suggests nurturing and providing for in addition to the intelligence and planning ordinarily associated with design. In a discussion of the distribution of plants, Somerville tells us, "Providence has endowed those most essential to man with a greater flexibility of structure, so that the limits of their production can be extended by culture beyond what have [sic] been assigned to them by nature." (PG 355) Other aspects of this notion of Providence are reflected by the following passage, which concludes a discussion of the role of earthworms in improving soil:

> Thus, the most feeble of living things is employed by Providence to accomplish the most important ends.... Harmony exists between the animal and vegetable creations; animals consume the oxygen of the atmosphere, which is restored by the exhalation of plants, while plants consume the carbonic acid exhaled by men and animals; the existence of each is thus due to their reciprocal dependence. Few of the great cosmical phenomena have only one end to fulfil, they are the ministers of the manifold designs of Providence. (PG 298–299)

In *Mechanism* and *Connexion*, the reader followed gravity through the mazes of the universe. In *Physical Geography*, the phenomena of interest are "the ministers of the manifold designs of Providence," which are forces and entities that are multiply connected and often transformed from one state into another.

The notion of reciprocal dependence is central to Somerville's thinking about physical geography, and the emphasis on the relationship
between human beings and nature that it entails is a new dimension
added to the theme of complexity within unity. She offers us a nuanced
and sophisticated view of the relationship between human beings and
nature, a view that insists on the right of human beings to control and
use nature but also recognizes the mutual influence of human beings
and the environment, or, to put it more precisely, that human beings
are part of the environment. This relationship ultimately does not seem
to be either adversarial or egalitarian but is instead an interesting
mixture of hierarchy and interdependence, which is in many ways
similar to Victorian notions of social organization. "Man himself is
viewed but as a fellow-inhabitant of the globe with other created
things, yet influencing them to a certain extent by his actions, and influenced in return." (PG 13) Somerville asserts the superiority of human
beings but balances this with emphasis on the interdependence and
interconnectedness of creation – human beings are superior but still
quite dependent, and much of the book is devoted to demonstrating
the myriad ways in which this is true.

In many cases, Somerville delineates the connections among and
transformations occurring in the earth, the sea, the air, and their inhabitants to such an extent that the distinction between animate and inanimate nature becomes blurred.

> The quantity of fossil remains is so great that . . . probably not a particle
> of matter exists on the surface of the earth that has not at some time
> formed part of a living creature. Since the commencement of animated
> existence, zoophytes have built coral reefs extending hundreds of miles,
> and mountains of limestone are full of their remains all over the globe
> . . . shells not larger than a grain of sand form entire mountains; a great
> portion of the hills of San Casciano, in Tuscany, consist of chambered
> shells so minute that Signor Soldani collected 10,454 of them from one
> ounce of stone. Chalk is often almost entirely composed of them. . . .
> there are even hills of great extent consisting of this substance, the débris
> of an infinite variety of microscopic insects. (PG 35)

She also reiterates one of her earlier themes – the notion of the
unchanging nature of God in the face of the vicissitudes of physical
existence – but emphasizes new sources of power and information in
nature.

> The earthquake and the torrent, the august and terrible ministers
> of Almighty Power, have torn the solid earth and opened the seals of
> the most ancient records of creation, written in indelible characters
> on the "perpetual hills and the everlasting mountains." There we read

of the changes that have brought the rude mass to its present state, and of the myriads of beings that have appeared on this mortal stage, have fulfilled their destinies, and have been swept from existence to make way for new races, which, in their turn, have vanished from the scene, till the creation of man completed the glorious work. Who shall define the periods of those mornings and evenings when God saw that his work was good? . . . Yet man is also to vanish in the ever-changing course of events. . . . where all is variable but the laws themselves, and He who has ordained them. (PG 14)

This passage describes an appreciative encounter with the destructive power of nature, "the earthquake and the torrent, the august and terrible ministers of Almighty Power." As the "seals of the most ancient records" are broken and geologists read the story of the earth, we return to the scene of creation in Genesis and see that the story did not end on the seventh day, that it has continued and will go on far beyond any scale of time that human beings are capable of grasping. The strong biblical allusion here might be taken to suggest that she is attempting to support a literal interpretation of the creation story in Genesis. What she is actually doing is mapping the geological theories of Charles Lyell onto the biblical account.

Somerville's identification with nature and with the cyclical processes she describes is striking. She views these in a positive light and indeed seems to exult in them. While many Victorians may have found the implications of geology depressing,[1] this was clearly not true for Somerville, largely because she identified so strongly with the Creator who directed the processes involved. Her work demonstrates that a belief in the superiority of man and tremendous pride in human accomplishments can be quite compatible with an acceptance of man's very small place in a much larger and more powerful scheme of things.

Another important theme in the interpretive frame of *Physical Geography* is the idea that the story told by the earth is different than that told by the heavens. Even though it takes place on an immense time scale, the story told by the earth has a beginning and suggests an end. The story of the heavens, in contrast, seems unbounded by time.

> The solid earth thus tells us of mountains washed down into the sea with their forests and inhabitants; of lands raised from the bottom of the ocean loaded with the accumulated spoils of centuries; of torrents of water and torrents of fire. In the ordinances of the heavens no voice

---

[1] See, for example, Dennis R. Dean, "Through Science to Despair: Geology and the Victorians," in *Victorian Science and Victorian Values*, ed. James Paradis and Thomas Postlewait (New York: New York Academy of Science, 1981), pp. 111–136.

declares a beginning, no sign points to an end; in the bosom of the earth, however the dawn of life appears, the time is obscurely marked when first living things moved in the waters, when the first plants clothed the land. (PG 33)

When Somerville recounts this story, it is conveyed as traces of evidence or snapshots captured in particular physical phenomena, rather than as a continuous narrative. She does not tell the story from beginning to end but rather offers us glimpses, enough to see the immensity of the forces and time scales involved but not enough for us to understand the entire unfolding. This is consistent with her desire to preserve some mystery associated with natural processes and to avoid positing a mechanism by which new life forms appear. Throughout *Physical Geography*, she emphasizes the immensity of the time scale involved to buttress one of her most important and consistent arguments: that the change from one species to another "never was abrupt . . . that there is no proof of progressive development of species by generation from a low to a high organization" (PG 34) and that change of species is impossible.

In the interpretive frame for *Physical Geography*, God as designer of intricate yet regular mazes is still present but overshadowed by images of providence, plenitude, myriad beings, great changes taking place over periods of time beyond calculation, and active forces on the earth beyond imagination. Referring to fossils of the Eocene period, she says, "This marvellous change of the creative power was not confined to the earth and the ocean; the air also was now occupied by many extinct races." (PG 29) She describes fossil remains as "innumerable," "inexhaustible multitudes," and "prodigious quantities." (PG 36) The mountains are formed by "repeated violent conclusions" through "incalculable . . . majestic operations." (PG 37) Alongside the violent convulsions are slower and gentler processes that eventually have enormous influence because of their pervasiveness and the very long periods of time over which they act. All of these phenomena support views of both gradual and dramatic change.

The picture of the earth created through physical geography as Somerville conceives it is of "the marvellous history laid open to us on the earth's surface. Surely it is not the heavens only that declare the glory of God – the earth also proclaims his handiwork!" (PG 36) Like other passages in *Physical Geography*, this one draws on the language of the Psalms, in this case Psalms 19:1, but we can gain further insight into the rhetoric Somerville is using from the footnote she places at the end of this statement. Instead of referring, as a contemporary reader might expect, to the source of the biblical quotation, she refers instead to notable geologists (Cuvier, Lyell, Murchison, de la Beche, and Owen) who interpret geology as being consistent with a universe governed by design and purpose. Her use of these authorities reveals both the ways in which scientific authority of the highest kind could be used to support theistic beliefs and the fact that the debate about the ultimate meaning of geological findings was widespread. The framework Somerville develops attempts to locate the story of the earth within the larger framework of the creator who had been the main character in the story of the heavens.

Descriptive Patterns

The interpretive frame of *Physical Geography* is developed, explored, and reinforced using several descriptive patterns. She continued to use concrete and easily grasped phenomena to provide insight to processes that take place at the limits of imagination. For example, she uses a discussion of the thickness of the strata of the earth as an illustration of the immensity of the time scales involved in geological processes. To develop the point, she says,

> Every river carries down mud, sand, or gravel, to the sea: the Ganges brings more than 700,000 cubic feet of mud every hour, the Yellow River in China 2,000,000, and the Mississippi still more; . . . if the sediment of all the rivers on the globe were spread equally over the bottom of the ocean, it would require 1,000 years to raise its bed one foot . . . so that the whole period from the beginning of the primary fossiliferous strata to the present day must be great beyond calculation, and only bears comparison with the astronomical cycles, as might naturally be expected, the earth being without doubt of the same antiquity with the other bodies of the solar system. (PG 33–34)

This strategy of using concrete and familiar phenomena as a base from which the reader could ascend to truths beyond the reach of ordinary human perception continues in this work to be one of the strengths of her style of presenting science.

The painterly qualities of Somerville's writing are more fully developed in *Physical Geography* than in her earlier works, in part because the subject matter consisted largely of landscapes of one kind or another, so that there is greater emphasis on color and light. In addition, there are two new descriptive patterns she uses, patterns that are distinct yet related and mutually reinforcing. The first, which I have designated "nature as epic theater," highlights contrasting, conflicting, or contesting forces. The other, which I have called "the living landscape," portrays the natural world as perpetually moving and changing, dying and renewed.

Light and Color

Although light and color were discussed in Somerville's first two works, they are exploited to much greater aesthetic effect in *Physical Geography*. In her description of the oceans, she includes the following passage on the colors of the ocean:

> The purest spring is not more limpid than the water of the ocean; it absorbs all the prismatic colours, except that of ultramarine, which, being reflected in every direction, imparts a hue approaching the azure of the sky. The colour of the sea varies with every gleam of sunshine or passing cloud, although its true tint is always the same when seen sheltered from atmospheric influence. The reflection of a boat on the shady side is often of the clearest blue, while the surface of the water exposed to the sun is bright as burnished gold. The waters of the ocean also derive their colour from animalcules of the infusorial kind, vegetable substances, and minute particles of matter. It is white in the Gulf of Guinea, black round the Maldives; off California the Vermilion Sea is so called on account of the red colour of the infusoria it contains. . . . Rapid transitions take place in the Arctic Sea, from ultramarine to olive-green, from purity to opacity. . . . the green is produced by myriads of minute insects, which devour one another and are a prey to the whale. The colour of clearer shallow water depends upon that of its bed; over chalk or white-sand it is apple-green, over yellow sand dark-green, brown or black over dark ground, and grey over mud. (PG 190–191)

This description not only paints a vivid picture of the ocean and all its manifestations of color but also makes very fine distinctions between colors and shades, and offers insight into the phenomena that create the various colors. She includes a similar passage on the colors of rivers. (PG 242)

As part of her account of North American flora, Somerville offers this description, which echoes her youthful responses to the autumn leaves she observed at Jedburgh:

> The autumnal tints of the forests of the middle States are beautiful and of endless variety; the dark leaves of the evergreen pine, the red foliage of the maple, the yellow beech, the scarlet oak, and purple Nyssa, with all their intermediate tints, ever changing with the light and distance, produce an effect at sunset that would astonish the native of a country with a more sober-coloured flora under a more cloudy sky. (PG 342)

The reader of this passage would scarcely guess that Somerville never saw this particular set of phenomena in person, since she never visited the United States. Numerous passages throughout *Physical Geography* attest to the power of Somerville's imagination and to her ability to

provide readers with the sense that they, too, had witnessed the phenomena she described.

Her discussion of lightning and electricity demonstrates the way in
which she was able to dramatize natural phenomena and to portray
them simultaneously in aesthetic, practical, and meditative modes.
Following a description of Professor Wheatstone's experiments
to measure the velocity of lightning, which revealed that it "would
encircle the globe in the twinkling of an eye," she says,

> This inconceivable velocity is beautifully exemplified in the electric tele-
> graph, by which the most violent and terrific agent in nature is rendered
> obedient to man, and conveys his thoughts as rapidly as they are formed.
> The colour of lightning is generally a dazzling white or blue, though if
> highly rarified it is rose-colour or violet. (PG 288)

The multiple appeals of electricity or lightning as exemplified in this
passage are those she sees throughout natural phenomena, and though
she may focus on one dimension at a time, she does not see them as in
any way being mutually exclusive or in competition – they are mutually reinforcing modes of appreciation and perception.

She also develops biblical themes in her treatment of light, specifically when she describes the solar spectrum as it affects plants, a subject
on which she had published experimental work. In a footnote to her
discussion on the nourishment of plants, she explains:

> The solar spectrum, or coloured image of the sun, formed by passing a
> sunbeam through a prism, is composed of a variety of invisible as well
> as visible rays. . . . There are also peculiar rays which produce phospho-
> rescence, others whose properties are not quite made out, and probably
> many undiscovered influences; for time has not yet fully revealed the
> sublimity of that creation, when God said, "Let there be light – and there
> was light." (PG 302)

In referring to Genesis, Somerville is not suggesting that her reader
accept a literal biblical account of creation in seven days (her remarks
on geological time scales make this clear). Still, she does seem to
be linking science to biblical traditions and evoking the sublime of
the physico-theological poets, especially as they were inspired by
Newton's *Opticks* and his theories on light.

Nature as Epic Theater

In this descriptive pattern, Somerville puts the reader in the position
of spectator or audience at the contest of conflicting forces in nature.
Sometimes the scene is explicitly cast as theater; in other cases, it is
implicit, and the reader simply observes a spectacle. Sometimes it is a

spectacle of richness, plenitude, and exuberance. Sometimes it is a land-scape of desolation. Either way, Somerville is able to create a sense of drama and evoke the sublime in her description of natural forces. In nature as epic theater, we see drama, spectacle, and conflict at both the microscopic and macroscopic levels.

Somerville clearly was not unique in portraying nature as epic theater, but she was particularly successful.[2] One region of Iceland she describes as seeming "to rest on an ocean of fire," with mountains that "conceal under a cold and tranquil coating of ice the fiery germs of terrific convulsions" (PG 161); at the extremities of this region are "theaters of perpetual volcanic activity." (PG 161) Between the mountains

> is a tremendous desert, never approached without dread, even by the natives – a scene of perpetual conflict between the antagonist powers of fire and frost, without a drop of water or a blade of grass; no living crea-ture is to be seen – not a bird, nor even an insect. The surface is a con-fused mass of streams of lava rent by crevices; and rocks piled on rocks, and occasional glaciers, complete the scene of desolation. (PG 161)

On the microscopic scale, she describes the process by which a seed develops into a plant and plants draw their sustenance from the earth and air. She portrays processes of change and transformation as the components of soil decompose and become nourishment for plants through a complex series of operations.

> It is wonderful that so small a quantity of carbonic acid as exists in the air should suffice to supply the whole vegetation of the world – and still more wonderful that a seed, minute enough to be wafted invisibly by a breath of air, should be the theater of all the chemical changes that make it germinate. (PG 300)

She also describes the "contest between spring and winter" in the steppes of Siberia. (PG 74)

In describing natural forces on a large scale, she portrays two cate-gories of sublime phenomena – the ones that move rapidly and dra-matically and the ones that move inexorably, though less dramatically,

---

[2] She includes a vivid example of another writer's use of nature as epic theater in her description of lagoon islands. She quotes a Mr. Jukes, a naturalist on the voyage of Captain Blackwood to explore the Torres Straits, as he describes the rolling of the waves along a coral reef in Australia. Jukes describes "one great continuous ridge of deep blue water, which curling over, fell on the edge of the reef in an unbroken cataract of daz-zling white foam" and concludes, "There was a simple grandeur and display of power and beauty in this scene that rose even to sublimity. . . . Both the sight and the sound were such as to impress the spectator with the consciousness of standing in the pres-ence of an overwhelming majesty and power." (PG 146)

and work remarkable changes over time. One example of this occurs as she portrays some of the drama that occurs beneath the earth:

> The power of the earthquake in raising and depressing the land has long been well known, but the gradual and almost imperceptible change of level through immense tracts of the globe is altogether a recent discovery; it has been ascribed to the expansion of rocks by heat, and subsequent contraction by the retreat of the melted matter from below them. It is not at all improbable that there may be motions, like tides, ebbing and flowing in the internal lava. (PG 158)

She echoes Genesis in her discussion of the nature and character of mineral veins.

> The tumultuous and sudden action of the volcano and the earthquake on the great masses of the earth is in strong contrast with the calm, silent operations on the minute atoms of matter by which Nature seems to have filled the fissures in the rocks with her precious gifts of metals and minerals, sought for by man from the earliest ages to the present day. Tubalcain was "the instructor of every artificer in brass and iron." (PG 168)

The reference to Tubal-cain[3] comes from Genesis 4:22. Somerville thus evokes traditional biblical accounts of creation even as she presents modern scientific ones.

The following passage on the discussion of the aurora shows the theme of nature as epic theater combined with vivid description of light with all its color and movement. She tells us that the aurora

> generally appears soon after sunset in the form of a luminous arch . . . across the arch the coruscations are rapid, vivid, and of various colours, darting like lightning to the zenith, and at the same time flitting laterally with incessant velocity. . . . they sometimes surpass the splendour of the stars of the first magnitude, and often exhibit colours of admirable transparency, blood-red at the base, emerald-green in the middle, and clear yellow towards their extremity. Sometimes one, and sometimes a quick succession of luminous currents run from one end of the arch or bow to the other, so that the rays rapidly increase in brightness. . . . The rays occasionally dart far past the zenith, vanish, suddenly reappear, and, being joined by others from the arch, form a magnificent corona or immense dome of light. (PG 290)

As illustrated earlier, Somerville also finds drama and the sublime in scenes of desolation. Her description of the outline of the land in Greenland conveys much the same sense that we get from Mary Shelley's descriptions of the frozen territory through which Victor

---

[3] Tubal-cain or Tubalcain was the third son of Lamech and Zillah, the first smith in the Old Testament.

Frankenstein pursues the creature. This description embodies the notion of counterpoise, or balancing weights, which was an important element in her aesthetics.

> Greenland . . . the domain of perpetual snow . . . seems to form a counterpoise to the preponderance of dry land in the northern hemisphere. There is something sublime in the contemplation of these lofty and unapproachable regions – the awful realm of ever-enduring ice and perpetual fire, whose year consists of one day and one night. The strange and terrible symmetry in the nature of the lands within the polar circles, whose limits are to us a blank, where the antagonist principles of cold and heat meet in their utmost intensity, fills the mind with the awe which arises from the idea of the unknown and the indefinite. (PG 38–39)

This passage uses a lofty and unapproachable setting to stage the conflict of the perpetual antagonisms of fire and ice, cold and heat. The knowledge on which this description is based comes from observation, not calculation. Though observation and measurement do not generate the kind of awe inspired by higher mathematical analysis, there is still awe to be perceived in the scene itself, especially when the science is presented with such vivid and evocative description. The writer's challenge is not to translate knowledge from symbols into verbal expressions but rather to recreate in the imagination remote scenes and processes. The remoteness may be literal (distance) or figurative (insight). In any case, Somerville shows herself again to be both a mathematician and a poet in soul, one who sees more than others see and who can share the vision.

The desert provides another stage for the sublime as she describes the scorching heat, the pinching cold, and the blasting hurricane winds in the Sahara.

> The desolation of this dreary waste, boundless to the eye as the ocean, is terrific and sublime; the dry heated air is like a red vapour, the setting sun seems to be volcanic fire, and at times the burning wind of the desert is the blast of death. . . . thick incrustations of dazzling salt cover the ground, and the particles, carried aloft by whirlwinds, flash in the sun like diamonds. (PG 90)

There is something reminiscent of the *Rime of the Ancient Mariner* in this passage, the same sense of expansiveness but also of color, movement, drama, and vivid description of all of these.

In her description of other deserts of Africa, those "cut transversely by the Nile," we find an excellent example of her ability to display contrasting elements to create both vividness and drama. She also emphasizes an element that she often used in her earlier works on astronomy – the association of silence with the sublime.

On the interminable sand and rocks of these deserts no animal – no insect – breaks the dread silence; not a tree nor a shrub is to be seen in this land without a shadow. In the glare of noon the air quivers with the heat reflected from the red sand, and in the night it is chilled under a clear sky sparkling with its host of stars. (PG 91)

Some of the most sublime locations are those on the earth but still far above the habitations of human beings and the daily experience of most readers. Somerville uses another common scene of the sublime, the mountains, as exemplified in this description of the Himalayas. In some ways, this passage also evokes the descriptive pattern of tracing the mazes.

The loftiest peaks being bare of snow give great variety of colour and beauty to the scenery, which in these passes is at all times magnificent. During the day, the stupendous size of the mountains, their interminable extent, the variety and sharpness of their forms, and, above all, the tender clearness of their distant outline melting into the pale blue sky, contrasted with the deep azure above, is described as a scene of wild and wonderful beauty. . . . There, far above the habitation of man, no living thing exists, no sound is heard; the very echo of the traveller's footsteps startles him in the awful solitude and silence that reigns in these august dwellings of everlasting snow. (PG 62–63)

Like her description of sound in the "Preliminary Dissertation," this passage allows the reader to move through a consideration of terrestrial phenomena and eventually to transcend ordinary experience.

The notion of nature as epic theater is equally well carried out in her description of the glaciers in the mountains of the Andes

where nature has been shaken by terrific convulsions. The dazzling snow fatigues the eye; the huge masses of bold rock, the mural precipices, and the chasms yawning into the dark unknown depths, strike the imagination; while the crash of the avalanche, or the rolling thunder of the volcano, startles the ear. (PG 102)

This passage, too, reflects an aesthetic of contrasting or conflicting experiences and forces, and of the sublime exhibited in silence and desolation. It is important to keep in mind that passages like these play a relatively small role in the book, since most of it presents details regarding physical phenomena – data, descriptions, and theories. The sublime and poetic ruminations are interwoven, do not appear at any particular intervals, and pass without additional comment from the author. Often, the subjects that might seem to be most particularly appropriate for poetic nature writing, flowers for example, receive relatively little treatment unless they are part of a grand landscape or dramatic scene.

Somerville takes her reader on a tour of the globe, often with a perspective from above. Although there is no cosmic platform from which to watch, there is something roughly equivalent. The reader views the volcanoes of Hawaii with their "awful sublimity" and observes "lakes of lava" and cauldrons "in furious ebullition . . . streams of lava, hurrying along in fiery waves . . . finally precipitated down an ignited arch, where force . . . literally spun them into threads of glass." (PG 151)

Somerville puts the reader in a position where a comprehensive view can be obtained. For example, she describes the relative distributions of land and ocean on the globe using the perspective of a person raised above the earth. She begins with specific facts relating to the relative distribution of land and ocean on earth and then says:

> A person raised above Falmouth, which is almost the central point, till he could perceive a complete hemisphere, would see the greatest possible expanse of land, while, were he elevated to the same height above New Zealand, he would see the greatest possible extent of ocean. (PG 38)

She follows with much more detailed information about the areas of the continents in square miles and their relative sizes.

Like all skillful, imaginative writers, Somerville is able to transport the reader to places he or she has never seen (and, amazingly enough, in most cases, that Somerville herself has never seen) and to inspire the reader's feeling for nature and nature's God. She thus exploits and in many cases greatly expands the full range of responses a human being might have when faced with these scenes. Yet she does this while at the same time conveying highly accurate observations and measurements of the phenomena under consideration and offering the latest scientific understanding of the forces at work and their history. She offers expertise-based descriptions that seem on the surface to be those of a painter or poet. Her work is an ongoing refutation of the belief that detailed scientific examination of nature eliminates aesthetic response.

The Living Landscape

The other descriptive pattern Somerville uses is the living landscape, which she portrays at both the microscopic and macroscopic levels. This pattern is particularly well suited to blurring the distinction between animate and inanimate nature and between human and animal life. Referring to both existing and extinct animals, she says:

> How wonderful the quantity of life that now is, and the myriads of being
> that have appeared and vanished! Dust has returned to dust through a
> long succession of ages, and has been continually remoulded into new
> forms of existence – not an atom has been annihilated. (PG 435)

This passage echoes both Genesis and the *Book of Common Prayer*.
In a footnote to the description quoted above, she paints an image
that recurs throughout *Physical Geography*, that of a living landscape,
perpetually dying and perpetually being renewed.

> The surface of the globe may be viewed as one mass of animal life – per-
> petually dying – perpetually renewed. A drop of stagnant water is a
> world within itself, an epitome of the earth and its successive geological
> races. A variety of microscopic creatures appear, and die; in a few days
> a new set succeeds; these vanish and give place to a third set, of differ-
> ent kinds from the preceding; and the débris of all remain at the bottom
> of the glass. (PG 435)

The notion of the microcosm – the drop of water as "a world within
itself" – became dominant in *On Molecular and Microscopic Science*.
Most of *Physical Geography* is devoted to the living landscape on a
grand scale.

She offers the following description of the flora of South America,
which emphasizes color, movement, and change, the metamorphosis
of one form of natural life into another:

> No language can describe the glory of the forests of the Amazon and
> Brazil, the endless variety of form, the contrasts of colour and size. . . .
> Passifloræ and slender creepers twine round the lower plants, while
> others as thick as cables climb the lofty trees, drop again to the ground,
> rise anew and stretch from bough to bough, wreathed with their own
> leaves and flowers, yet intermixed with the vividly coloured blossoms
> of the Orchideæ. An impenetrable and everlasting vegetation covers the
> ground; decay and death are concealed by the exuberance of life; the trees
> are loaded with parasites while alive – they become masses of living
> plants when they die. (PG 348)

The animal life of the Amazon forests are equally lively and equally
vividly portrayed, as in this passage which gives particular emphasis
to sound:

> The beasts seem to be periodically and unanimously roused by some
> unknown impulse, till the forest rings in universal uproar. Profound
> silence prevails at midnight, which is broken at the dawn of morning by
> another general roar of the wild chorus. Nightingales too have their fits
> of silence and song; after a pause they

"– all burst forth in choral minstrelsy,
As if some sudden gale had swept at once
A hundred airy harps."
                                        Coleridge

The whole forest often resounds when the animals, startled from their
sleep, scream in terror at the noise made by bands of its inhabitants flying
from some night-prowling foe. Their anxiety and terror before a thunderstorm is excessive, and all nature seems to partake in the dread. (PG 108)

The strong portrayal of affective response in the animals of the jungle
helps elicit a similar response in a human reader.

One of the most striking features of the pattern of the living landscape is the emphasis it places on the movement, real or apparent,
of the landscape, such as creeping plants that clothe mountains and
pastures that undulate. (PG 105) She also focuses on natural processes
that appear magical, as in these descriptions of Siberia and the area
near the Cape of Good Hope:

The scorching heat of the summer's sun produces a change like magic
on the southern provinces of the Siberian wilderness. The snow is
scarcely gone before the ground is covered with verdure, and flowers of
various hues blossom, bear their seeds, and die in a few months, when
winter resumes his empire. (PG 76)

In a later passage, she reports that "the sudden effect of rain on the
parched ground is like magic: it is recalled to life, and in a short time
is decked with a beautiful and peculiar vegetation." (PG 333) This is a
landscape that is not just alive but moving, changing, interacting, transforming constantly and often in dramatic ways. These passages
provide evidence of the excellent fit that existed between Somerville's
interpretive and literary skills and the nature of the phenomena she
portrayed.

Mary Somerville on the Human Condition

The last chapter of *Physical Geography*, which is entitled "The Distribution, Condition, and Future Prospects of the Human Race," deals with
what might be considered the most active element in the living landscape – human beings. In her discussion of humanity, Somerville covers
a very wide range of topics of great interest to Victorians, from the
various races of human beings, their characters and distributions, the
factors that influence human character and development to human
inventions, emigration and colonization, the advancement of science,
literature, and painting. Since a complete analysis of her thoughts on

all of these topics and her relation to her contemporaries would constitute a complete study of its own, the discussion here only touches lightly on some of the most interesting and those that are most directly related to science: the permanency of species; the balance of power between human beings and nature; the influence of the natural and social environments on human character and behavior; and individuality, civilization, and progress, with particular emphasis on the multiple roles that science and technology play in human advancement.

Her views are in many ways typical of her age, in some ways divergent, with a strong tendency toward Romanticism and a strong belief in progress. One of the most striking characteristics of her views is that they tend not to be deterministic or reductionist. She believes that the conditions we observe in both the human and non-human realms are determined by complexes of factors, not single, dominating ones; although she feels she can safely generalize about whole races or nations of people, she also recognizes individual differences and variations: "the difference in the capabilities of nations seems to be as great as that of individuals." (PG 437)

On the issue of whether all human beings are part of the same species, she concluded that we are indeed a single species. Though she saw Europeans as the most advanced race and tended to see American Indians as the least advanced, she found something to praise about every race she described: the Caucasian, Mongol-Tartar, Malayan, Ethiopian, and American. In her view, natural causes, such as a mild climate or the availability of rivers to facilitate the exchange of goods and ideas, combined with mental factors such as superior intelligence or initiative to lead to different lines of development.

All of humanity is fundamentally linked by what she refers to as "the immortal spirit," which is "the chief agent in all that concerns the human race." This concept, which she calls "the universal mind" (PG 436), seems very much related to but is not necessarily identical with the human soul.

> The progress of the universal mind in past ages, its present condition, and the future prospects of humanity, rouse the deep sympathies of our nature, for the high but mysterious destiny of the myriads of beings yet to come, who, like ourselves, will be subject for a few brief years to the joys and sorrows of this transient state, and fellow-heirs of eternal life hereafter. (PG 436)

Thus, all human beings are linked by mortality and immortality and by their various forms of interdependence.

She in turn sees human beings as linked to and on a continuum with other living beings, while at the same time being much more complex and generally superior. One of the most important sources of human superiority is the flexibility that allows human beings to adapt to varied circumstances of climate and diet. These varied circumstances contribute substantially to the physical variability of humankind: "The arrangement of Divine Wisdom is to be admired in this as in all other things, for, if man had only been capable of living on vegetable food, he never could have had a permanent residence beyond the latitude where corn ripens." (PG 446)

One of her most important philosophical commitments is to the permanency of species, including the human species. The influence of the environment that she emphasizes so much and points to in so many cases supports this position by accounting for the great variability in human beings: "Every change of food, climate, and mental excitement, must have their influence on the reproduction of the mortal frame; and thus a thousand causes may co-operate to alter whole races of mankind placed under new circumstances, time being granted." (447) Within this view of things, the experience of manual labor as opposed to mental labor accounts for differences in the countenances of individuals: "The refining effects of high culture, and, above all, the Christian religion, by subduing the evil passions, and encouraging the good, are more than anything calculated to improve even the external appearance." (PG 447)

This line of argument builds to one of the most important points she makes, one that she had insisted upon in less explicit guises throughout *Physical Geography*:

> Thus, an infinite assemblage of causes may be assigned as having produced the endless varieties in the human race; but the fact remains an inscrutable mystery. But amidst all the physical vicissitudes man has undergone, the species remains permanent; and let those who think that the difference in the species of animals and vegetables arises from diversity of conditions, consider, that no circumstances whatever can degrade the form of man to that of the monkey – or elevate the monkey to the form of man. (PG 447)

Later in the same discussion, she refines this argument with a theme she used earlier – stability within vicissitudes.

> The circumstances which thus determine the location of nations, and the fusion or separation of their languages, must, conjointly with moral causes, operate powerfully on their character. The minds of mankind, as well as their fate, are influenced by the soil on which they are born and bred. . . . Early associations never entirely leave us, however much our position in life may alter, and strong attachments are formed to places which generate in us habits differing from those of other countries. (PG 449)

She believes that "among the objects which tend to the improvement of our race, the flower-garden and the park adorned with native and foreign trees have no small share." (PG 452) Her work provides ample evidence of the many positive feelings and values that could be associated with gardens and parks and of the kind of influence that they might have.

Somerville is very much committed to a progressive view in which science and technology play a large role:

> The influence of external circumstances on man is not greater than this influence on the material world. He cannot create power, it is true, but he dexterously avails himself of the powers of nature to subdue nature. Air, fire, water, steam, gravitation, his own muscular strength, and that of animals rendered obedient to his will, are the instruments by which he has converted the desert into a garden, drained marshes, cut canals, made roads, turns [sic] the course of rivers, cleared away forests in one country, and planted them in another. (PG 451)

Despite her enthusiasm for these advances, she also recognized a destructive propensity in humanity and urges a balanced approach.

> Man's necessities and pleasure have been the cause of great changes in the animal creation – and his destructive propensity of still greater. Animals are intended for our use, and field-sports are advantageous by encouraging a daring and active spirit in young men, but the utter destruction of some races, in order to protect those destined for his pleasure, is too selfish. (PG 453)

She worries that "The lion, the tiger, and the elephant will be known only by ancient records" (PG 453) as their habitats are destroyed by human development. She does not, however, believe that human beings can dominate nature in the long run: "Man, the lord of the creation, will extirpate the noble creatures of the earth – but he himself will ever be the slave of the canker-worm and the fly. . . . the insect tribe [who] will come from the desert and destroy the fairest prospects of the harvest." (PG 453–454) In the picture she paints, humanity is clearly quite influential but certainly not able to transcend or control all aspects of nature.

Despite the influence of the environment on man, and of man on the natural environment, the most important processes in progress are human and social, collective not individualistic.

> The influence of man on man is a power of the highest order, far surpassing that which he possesses over inanimate or animal nature. It is, however, as a collective body, and not as an individual, that he exercises this influence over his fellow-creatures. The free-will of man, nay, even

his most capricious passions, neutralize each other, when large numbers of men are considered. (PG 454)

She traces the process by which the aggregated individuality "becomes the directrice of the principal social movements of a nation" and "private morality becomes the base of public morality." (PG 457)

It is not surprising that she sees discovery and innovation as collective social processes, but she also recognizes individual genius and effort.

> Scientific discoveries and social combinations, which put in practice great social principles, are not without a decided influence; but these causes of action coming from man, are placed out of the sphere of the free-will of each: so that individual impulse has less to do with the progress of mankind than is generally believed. When society has arrived at a certain point of advancement, certain discoveries will naturally be made; the general mind is directed that way, and if one individual does not hit upon the discovery, another will.... particular inventions or discoveries ... properly belong to the age in which they are made, without derogating from the merits of those benefactors of mankind who have lessened his toil or increased his comfort by the efforts of their genius. The time had come for the invention of printing, and printing was invented; and the same observation is applicable to many objects in the physical, as well as to the moral world. (PG 455–456)

The power of the collective mind is one of the things that makes Somerville so sure that the progressive tendencies of her age will continue unabated. Progress has many manifestations in a variety of realms, all of which are interconnected.

Although she believes that colonization is essentially progressive because it increases communication, stimulates commerce, and spreads civilization, she recognizes that it is not uniformly successful. It may be "the design of Providence to supplant the savage by civilized man" (PG 459), but she believes that the natives, "the ancient proprietors of the land," have rights, at least the very limited right not to be "swept away" (PG 458) or extinguished as they had been nearly in North America and completely in the West Indian islands. She is an ardent opponent of slavery, a topic raised from time to time in her survey of the globe. In Africa, "civilized man has been a scourge" and "extended its baneful influence into the heart of the continent, by the encouragement it has given to warfare among the natives for the capture of slaves, and for the introduction of European vices, unredeemed by Christian virtues." (PG 459) She sees hopeful signs in the resistance to slavery and believes that it will eventually be abolished.

Her greatest enthusiasms and greatest hopes for the future progress
of the human race arise from the advancement of science and technology and the spread of Christianity, which she sees as being related in
interesting ways. She sees science as an international and collective
enterprise, and a unifying factor.

> Science has never been so extensively and so successfully cultivated as
> at the present time: the collective wisdom and experience of Europe and
> the United States of America is now brought to bear on subjects of the
> highest importance in annual meetings, where the common pursuit of
> truth is as beneficial to the moral as to the intellectual character, and the
> noble objects of investigation are no longer confined to a philosophic few,
> but are becoming widely diffused among all ranks of society, and the
> most enlightened governments have given their support to measures that
> could not have been otherwise accomplished. (PG 462)

There is also evidence of progress in new discoveries:

> The places of the nebulæ and fixed stars, and their motions, are known
> with unexampled precision, and the most refined analyses embrace the
> most varied objects. Three new satellites and six new planets have been
> discovered within four years, and one of these under circumstances the
> most unprecedented. (PG 464–465)

But perhaps her most unbridled enthusiasm is for technological
developments, and her discussion of them reveals her gift for evoking
the technological sublime. The most important source of her enthusiasm for technology is its capacity to unite people in thought and across
distances. It helps link human being to human being.

> The history of former ages exhibits nothing to be compared with the
> mental activity of the present. Steam, which annihilates time and space,
> fills mankind with schemes for advantage or defence: but however mer-
> cenary the motives for enterprise may be, it is instrumental in bringing
> nations together. (PG 461)

Improved communication has provided greater incentives for people
to learn modern languages and facilitated the assimilation of character
amongst nations: "On earth, though hundreds of miles apart, that
invisible messenger, electricity, instantaneously conveys the thoughts
of the invisible spirit of man to man – results of science sublimely tran-
scendental." (PG 465) Like other encounters with the sublime, these
developments are almost beyond the reach of words to describe.

> Vain would be the attempt to enumerate the improvements in machin-
> ery and mechanics, the canals and railroads that have been made, the
> harbours that have been improved, the land that has been drained, the
> bridges that have been constructed; and now, although Britain is inferior

to none in many things, and superior to all in some, one of our most distinguished engineers declares that we are scarcely beyond the threshold in improvement; to stand still is to retrograde, human ingenuity will always keep pace with the unforeseen, the increasing wants of the age. (PG 465)

The limitless potential and the undeniability of the forces involved are keys to the sublime quality of technological progress, which moves through history much the way that gravity moves through the mazes of nature: "It would be difficult to follow the rapid course of discovery through the complicated mazes of magnetism and electricity." (PG 465)

In areas such as chemistry, physiology, psychology, geography, and literature, there have been similar advances, in a succession so rapid that "it is impossible to convey." (PG 465) She also sees evidence of great moral improvement and believes that the world is on the verge of "one of those important changes in the minds of men which occur from time to time, and form great epochs in the history of the human race." (PG 469)

> The present state of transition has been imperceptibly in progress, aided by many concurring circumstances, among which the increasing intelligence of the lower orders, and steam travelling, have been the most efficient. The latter has assisted eminently in the diffusion of knowledge. . . . No invention that has been made for ages has so levelling a tendency, which accords but too well with the present disposition of the people. The spirit of emancipation, so peculiarly characteristic of this century, appears in all the relations of life, political and social. (PG 470)

She sees these relations as characterized by the same mixture of hierarchy and interdependency that she saw in the natural world. Although she believes "inequality of condition is permanent as the human race," she also believes that "from necessity we must fulfil the duties of the station in which we are placed, bearing in mind that, while Christianity requires the poor to endure their lot with patience, it imposes a heavy responsibility on the rich." (PG 470–471) This is both an important form of social philosophy and the model on which Somerville organizes her view of the natural world. If we press the analogy to her description of nature, we can see that the lower orders possess considerable power despite their low place in the hierarchy.

Somerville concludes with what might be called a paean to progress in the nineteenth century; the signs of progress are as numerous as the features of nature to be observed.

> The moral disposition of the age appears in the refinement of conversation. Selfishness and evil passions may possibly be ever found in the

human breast, but the progress of the race will consist in the increasing power of public opinion, the collective voice of mankind regulated by the Christian principles of morality and justice. . . . there will probably always be a difference of view as to religious doctrine, which, however, will become more spiritual, and freer from the taint of human infirmity; but the power of the Christian religion will appear in purer conduct, and in the more general practice of mutual forbearance, charity, and love. (PG 474)

Thus, the plenitude, Providence, and progress that Somerville witnesses in the world of animate and inanimate nature, she sees mirrored in the tide of human affairs. She presents a vision of integrated progress of the earth, the sea, the air, and their inhabitants, a progress not without costs and failures but inevitable and universal nonetheless.

Viewing Somerville's Final Two Scientific Books in the Context of Her Career as a Whole

As noted earlier, Somerville began work on *Physical Geography* shortly after her departure for Italy and published it ten years afterward in 1848. Given these changes, it is interesting to consider how Somerville and her work were perceived. Henry Holland's review of *Physical Geography*, which was published in the *Quarterly Review* in 1848, provides some insight into both the character and quality of *Physical Geography* as an individual work and the way it was viewed as part of Somerville's total oeuvre.

Although Holland expresses a high opinion of Somerville as a scientist and a writer, he approaches the review in a critical spirit. Much of his criticism is of the type expected from a reader who is thoroughly familiar with the subject. For example, he contends that she does not devote enough attention to interesting new research on the interaction between the earth and its atmosphere (335) and notes "a few trifling errors" (324), which he feels sure will be corrected in subsequent editions. His most substantial criticisms concern instances where he believes Somerville has provided too much detail, especially "numerical facts" (322) without adequate generalizations to control them and thus has unduly taxed the reader's memory. This criticism can be attributed in part to his perception of the book as "an elementary work" (308), which it may not have been intended to be. Holland's perception of the need for more controlling generalizations may reflect the newness of the field of physical geography or Somerville's newness to the subject, but also may arise from the fact that *Physical Geography* has an interpretive framework but no preliminary dissertation. That is, it gives no sense of the historical process by which the discipline has developed and offers only very general treatment of the guiding principles behind it. Holland accurately asserts that Somerville simply "enters . . . abruptly on her theme." (310)

These reservations aside, Holland clearly sees *Physical Geography* as an important contribution to the field. In fact, he describes it as "the first English work bearing the title, and distinctly comprehending what belongs to [the] great subject" of physical geography. (306) In other words, he sees it as the first work in English to treat physical geography "as an independent branch of science." (306) His definition of the field strongly echoes Somerville's, especially in his insistence that it "excludes . . . all those artificial lines and names with which man has covered the earth" and that it includes "the earth, air, and ocean, with everything of animal and vegetable life tenanting these great domains." (309) In many ways, his praise of *Physical Geography* echoes Whewell's praise of *Connexion*. Like Whewell, he is distressed at the specialization and division into increasingly smaller departments of study that seems prerequisite for the advancement of science. He concludes reluctantly, "We divide, to obtain supremacy over the whole." (306) He sees physical geography generally and Somerville's treatment of it as a welcome corrective to this trend because it is "spacious" in "its domain and objects" and represents a move toward synthesis rather than division: "its subjects and subordinate branches, heretofore pursued under these separate connexions, [are now] associated under one comprehensive name." (306)

Throughout the review, Holland portrays Somerville in terms of expertise and innovation. Like Whewell, he sees Somerville as delineating and defining the conceptual landscape of science through *Physical Geography* and her earlier works. He admires her work on Laplace for the "mathematical capacity and cultivation, which enabled her to present to English readers an admirable summary of the spirit, methods, and results of this great work" (306–307), but reserves his greatest praise for *Connexion*, which he describes as "unassuming in form and pretensions, but so original in design and perfect in execution, as well to merit the success of eight editions, each carefully embodying all of augmentation that science had intermediately received." (307) Like *Physical Geography*, *Connexion* was in Holland's view a groundbreaking work:

Though rich in works on particular sciences, and richer still in those eminent discoveries which establish the relations amongst them, yet had we not before in English a book professedly undertaking to expound these connexions, which form the greatest attainment of present science, and the most assured augury of higher knowledge beyond. Mrs. Somerville held this conception steadily before her; and admirably fulfilled it. (307)

This assessment gives a sense of the magnitude of accomplishment *Connexion* represented for Somerville as an individual and for the scientific community, as well as of Somerville's steadfastness in pursuing the role and cultivating the distinctive space she had carved out for herself.

Holland further develops the notion of Somerville as innovator by arguing that *Connexion* provided the model for Humboldt's *Kosmos*:

> Her work indeed, though small in size, is a true Kosmos in the nature of its design, and in the multitude of materials collected and condensed into the history it affords of the physical phenomena of the universe. In some respects her scheme of treating these topics so far resembles that since adopted by Humboldt, that we may give Mrs. Somerville credit for partial priority of design. (307)

Quick to protect Somerville's reputation for modesty, he adds, "She would be the last person to assert it for herself." (307)

Holland also offers a comprehensive description and evaluation of "her style in treating scientific subjects" (307), which he characterizes in terms of "natural eloquence" and a preference for insight over ornament. This assessment is based on all three of the books she had written to that point and reads like an inventory of all the positive traits valued in her writing. It conveys the range of techniques she used for getting above Whewell's "cloud of words" in writing about science. These techniques include conciseness, which Holland attributes to "that native simplicity which is a quality of genius." (307) She avoids prefaces and perorations and "goes at once to the work in hand; fully prepared and informed; clear and exact in her methods; and always preferring perspicuity [insight] to ornament." (307)

In developing connections and "conducting" her reader to generalizations, she avoids "pomp of announcement" and substitutes instead "those clear and certain steps of induction which, better than any artifices of language, raise the mind to the height of the subject, and engage the imagination with visions of higher knowledge yet to come." (307) Holland also reflects the kind of authorial stance she establishes with respect to her subject:

> When writing on astronomy she allows the stars to speak for themselves, in all of their sublimities of number, space, and time; not defacing the history of the heavens by those gorgeous epithets which we find in some

modern treatises – words of earthy origin, and which rather debase than elevate the grandeur of the theme. (307)

Echoing many other assessments of Somerville's work, Holland emphasizes both its power and its restraint, "a character perfectly compatible with great merits of style, and passages of much natural eloquence." (307)

This restrained quality is not necessarily associated with the capacity for vivid description that is displayed most strongly in *Physical Geography* and *Molecular and Microscopic Science*, where the reader gets a particularly good view of Somerville's ability to dramatize nature and help the reader visualize remote or invisible processes. This ability allowed her to reach broader audiences than she had reached in her earlier works. Thus, it also raises questions about whether she might have deliberately adopted the less "professional" or less "scientific" authorial voice often associated with the traditions of natural history or nature writing. The less professional voice might be categorized as that of the "much-impressed witness" of the dramatic conflict of natural phenomena. This much-impressed witness has a strong narrative persona and is exemplified in Chapone's rhapsodic accounts of "the sensation I felt from the glorious, boundless prospect of infinite beneficence bursting at once upon my imagination!" (Chapone 1772: 147). Somerville was exposed to Chapone, and likely to similar authors, in her girlhood.

Somerville's authorial voice is clearly distinguishable from that of the much-impressed witness. As Michael Dettelbach says of von Humboldt's writing: "The forces themselves supply the narrative and the drama, and the story they tell is one of dynamic balance, equilibrium, and progress." (304)[4] The net result of such a strategy is a combination of power with restraint. I believe that one of Somerville's greatest innovations as a scientist was the style she developed for the presentation of science, which achieved natural eloquence by preferring insight over ornament. It thus blended the ideals and illuminating potential of modern science with the merits of the style associated with the tradition of the classical sublime and avoided the excesses of the much-impressed witness.

Physical Geography also offers perhaps the clearest view we get of Somerville in her role as organic intellectual and as possessor of both scientific and moral authority. Holland makes this clear when he talks

---

[4] Dettelbach is concerned only with what he calls "Humboldtian" science and does not discuss Somerville (or Chapone); his analysis draws on the work of Mary Louise Pratt, specifically *Imperial Eyes: Travel Writing and Transculturation* (London and New York: Routledge, 1992).

about the last sections of *Physical Geography*, those that deal with the human race and its prospects, a topic that Holland sees as: "a theme full of wonder and interest, pride and humiliation – painful in many points of view, perplexing and mysterious in all." (339) The need for guidance in thinking about the topic is intensified by "the time in which are now living" and the "new and mightier powers which man has formed for himself from the physical elements surrounding him." (339) The changes brought about in large part by science and technology leave "all old institutions and usages wrecked on the shore of an uncertain futurity." (339–340) Holland's closing remarks make it clear that Somerville's status as one of the scientific elite has also given her a kind of moral authority: "Mrs. Somerville's observations on this last great topic of her work are marked by the strong sense, clear discrimination, and warm and sincere piety which characterise her mind; and we earnestly recommend them, – as we again do all her writings, – to the careful study of our readers." (340)

Beyond the Power of Vision: On Molecular and Microscopic Science

On Molecular and Microscopic Science[5] was Somerville's last major scientific work, begun in her seventy-ninth year and published in her eighty-ninth. While it was arguably her least successful scientific book, it was in many respects a remarkable achievement, and its content and fate provide valuable insights into the intellectual history of the period. Many of the difficulties Somerville faced arose not from her advanced age or remoteness from research libraries but from the transformations in society and culture that were taking place at the time.

The research that she had undertaken for *Physical Geography* had led her to the conviction that the new frontiers of science were not in the heavens or the farthest corners of the earth, but in the realm of the infinitesimally small – the world of molecules and microbes that pulsed and whirled around and through us, largely without our even noticing it. In *On Molecular and Microscopic Science*, Somerville sought to provide access to the view from these frontiers. Neither she nor the reader would be disappointed at the view, which confirmed her assertion that

---

[5] The book consists of three parts in two volumes. Volume one contains parts one and two; Volume two contains part three. Volume one is referred to in the text with the abbreviation "MMS, 1," volume 2 with the abbreviation "MMS, 2."

"the deeper the research, the more does the inexpressible perfection of God's works appear, whether in the majesty of the heavens, or in the infinitesimal beings of the earth." (MMS, 2: 178) Like all frontiers, however, the new realms of molecular and microscopic science contained large tracts of contested territory.

One of the hallmarks of Somerville's works was their accuracy and currency. As William Whewell put it in his review of *Connexion*, "the writer had read up until Saturday night" (Whewell 1834: 560), and the frequent new editions Somerville produced of *Connexion* and *Physical Geography* reflected her commitment to providing an up-to-date view of her subject. This was particularly difficult to do in *On Molecular and Microscopic Science*, in large part because so much was changing in the fields she covered. The book nonetheless demonstrated that she had retained her capacity for coherent synthesis and for vivid and moving description, even with rather challenging material. It also provides insight into some of her most interesting philosophical commitments.

Numerous passages in *Physical Geography*, especially those dealing with the theme of the living landscape, reveal the extent to which the microscopic world engaged Somerville's imagination. She speaks of the inconceivable smallness of organisms and particles of matter. She sees a drop of water as "a world within itself" (PG 435), a grain of sand as the habitation of millions of minute organized beings (PG 280), and the ocean and the soil teeming with "microscopic live animals" (PG 368–369).

> This lowest order of animal life is much more abundant than any other, and new species are found every day. Magnified, some of them seem to consist of a transparent vesicle, and some have a tail: they move with great alacrity, and show intelligence by avoiding obstacles in their course: others have siliceous shells. Language, and even imagination, fails in the attempt to describe the inconceivable myriads of these invisible inhabitants of the ocean, the air, and the earth. (PG 368)

Similarly, in the inanimate world, "the smallest globule of steam tells no more of its atoms than the ocean; the minutest grain of sand magnified appears like the fragment of a rock – no mechanical division can arrive at the indivisible." (PG 280) She focuses on the inconceivable smallness of particles of matter whose nature and activity "are beyond the power of vision." (PG 280) One of her accomplishments in *On Molecular and Microscopic Science* was to get beyond the failures and limitations of language and vision and to use both language and science to create a vision that would expand the imagination so that it could more fully grasp the realm of the infinitely small.

In *On Molecular and Microscopic Science*, Somerville turned her attention to this world, the world of microscopically small forces and organisms, the world being progressively revealed through the microscope and through various forms of chemical and physical analysis in research programs carried out primarily by chemists, physicists, and biologists. She saw the "microscopic investigation of organic and inorganic matter [as] peculiarly characteristic of the actual state of science" (MMS, 1: preface n.p.) and set out "to give a sketch of some of the most prominent discoveries in the life and structure of the lower vegetable and marine animals in addition to a few of those regarding inert matter." (MMS, 1: preface n.p.) She organized and synthesized an enormous amount of detailed research regarding the structure and behavior of matter and very small organisms, and refers often to the work of a wide range of researchers. In *On Molecular and Microscopic Science*, as the writer for the *Edinburgh Review* expressed it,

> The gift of lucid description, so characteristic of the distinguished authoress of the "Connexion of the Physical Sciences," is as conspicuous as ever; but that which most forcibly strikes the reader of these pages is the extraordinary power of mental assimilation of scientific facts and theories which Mrs. Somerville displays. (Roscoe 1869: 137)

The book consists of three parts in two volumes. Part one deals with "Atoms and Molecules of Matter," Part two with "Vegetable Organisms," and Part three with "Animal Life." In her discussion of atoms and molecules, her theme is "the relation between the powers of nature and the particles of matter." (MMS, 1: 167) In parts two and three, her theme is "the study of the indefinitely small in the vegetable and animal creation." (MMS, 1: 167) She surveys the latest discoveries in fields including inorganic and organic chemistry, molecular physics, and physiology.

From an aesthetic point of view, the two volumes are quite elegant. They are nicely bound, and there are 180 illustrations, ten of them bold, full-page prints of cream on a striking cobalt blue background. As John Herschel's note congratulating Somerville on her most recent publication suggested, *On Molecular and Microscopic Science* reached a much broader audience than had *Mechanism* or even *Connexion*. "I have no doubt," Herschel writes on January 5, 1869, "it will prove very popular work and be found on many a drawing room table – accompanied I dare say with microscopes and other appliances for actually inspecting the wonderful creatures described." (Somerville Collection, MSH-3 in Dep.c.370) *On Molecular and Microscopic Science* brought readers closer to science so they could experience it first-hand.

The investigations which have revealed the most refined and wonderful relations between light, heat, electricity, and highly elastic media; the relation of these powers to the particles of solid and liquid matter, new methods of analysis, and the microscopic examination of that marvellous creation, animal and vegetable, which is invisible to the unaided eye of man, have brought a new accession to the indefinitely small within the limits of modern science. (MMS, 1: 1)

As it had in the celestial and terrestrial realms, science was bringing within the realm of human perception things that would otherwise have been far beyond it, and Somerville was making the view provided by science both real and accessible to her readers.

To present this view, Somerville once again used the pattern of nature as epic theater. In volume one, she uses Faraday's words and images to develop the theme. Faraday contrasts "the imposing grandeur" of the "majestic phenomena of combustion" with "the silent, tranquil, ever progressing metamorphic changes" such as decay.

When the loud crash of the thunder or the lightning's flash awakens us from our thoughtless abstractions or our reveries, our feelings become impressed with the grandeur of Omnipotence and the might of the elements he wields, yet the whole fury of the thunderstorm – what is that in comparison with the electric energies which silently and continually exert themselves in every chemical change? Why the electrical force in a single drop of water, and disturbed when the water is decomposed, is of itself greater than in the electricity of a whole thunderstorm. (MMS, 1: 10–11)

Faraday laments the tendency to concentrate on rapid, large-scale phenomena and to overlook "the aggregate results accomplished by [smaller scale forces] in the economy of the world." (MMS, 1: 11) Somerville revealed the drama in those forces and processes.

A new and unseen creation was brought under mortal eye, so varied, so astonishing, and inexhaustible, that no limit can be assigned to it. This invisible creation teems in the earth, in the air, and in the waters, innumerable as sand on the seashore. These beings have a beauty of their own, and are adorned and finished with as much care as the creatures of a higher order. (MMS, 1: 178)

This was truly the sense in which God seemed greatest in the least aspects of creation.

The reader also experiences nature as epic theater combined with the living landscape in the following passage, which follows a discussion of the cycles of existence in the vegetable realm. Somerville presents a vision of the highest grades of animal and vegetable life, "the monarchs of the forest" (MMS, 1: 178) falling and being destroyed through decay by the lower forms, "the moss and the lichen, resume their empire and life at the expense of the dying and the dead, a cycle which perpetuates the green mantle of the earth." (MMS, 1: 178) This image captures the drama of the forces at work in nature. The analogy between the realms revealed through the microscope and telescope is exploited in the illustrations included in the book. These illustrations are particularly dramatic and invite comparisons between the microscopic and celestial realms. Many of the microscopic organisms resemble planets or constellations of stars. When Somerville draws attention to these similarities, the effect is nothing short of spectacular, as in the following description of reef-building corals and the atolls and other structures they create. The microscopic realm seems to reveal the sublime most easily when it is metaphorically linked to the celestial realm.

> An atoll is a ring or chaplet of coral, enclosing a lagoon, or portion of the ocean, in its centre. . . . All the coral in the exterior of the ring, to a moderate depth below the surface of the water, is alive; all above it is dead, being the detritus of the living part washed up by the surf. . . . The waves give innate vigour to the polypes by bringing an ever-renewed supply of food to nourish them, and oxygen to aërate their juices; besides, uncommon energy is given and maintained by the heat of a tropical sun, which gives them power to abstract enormous quantities of solid matter from the water to build their stony homes, a power that is efficient in proportion to the energy of the breakers which furnish the supply. . . . In these calm, limpid waters the corals are of the most varied and delicate structures, of the most charming and dazzling hues. When the shades of evening come on, the lagoon shines like the Milky Way with myriads of brilliant sparks. The microscopic medusæ and crustaceans, invisible by day, form the beauty of the night, and the sea-feather, vermilion in daylight, now waves with green phosphorescent light. (MMS, 2: 143)

This passage allows the reader to see the impact of the accumulation of vast numbers of tiny living creatures in action. It combines nature as epic theater with the living landscape.

The journey into the microscopic world thus takes the reader into expanses of space and time and into a universe characterized by multiple interconnections. Molecular science demonstrates that the same chemical components exist on earth and in distant nebulae, the same gases on earth and in "distant regions of space." (MMS, 2: 249) Like many features of the inanimate landscape, microscopic organisms have remarkable endurance, the capacity to endure the vicissitudes that affect larger organisms: "At the most remote period of the sun's existence, the vivifying influence of the sun, the constitution and motions of the atmosphere and ocean, and the vicissitudes of day and night, of life and death, were the same as at the present time." (MMS, 2: 250) Microscopic examination of the physical world takes us to "remoteness in time. . . . far beyond the reach of imagination." (MMS, 2: 250) And in large part, this seems to be the primary aim of her work – to take the reader beyond the reach of imagination, by expanding the reader's imagination and leaving it at a frontier where there are still mysteries and wonder.

The large number of illustrations represent an innovation in Somerville's style of presentation, which typically used few diagrams and even fewer illustrations of other kinds. While the large blue-and-cream illustrations have an abstract quality that stimulates imagination, the same could probably not be said of the remaining 170 black-and-white illustrations, which seem more suitable for clinical study than aesthetic appreciation. The illustrations were expensive, and John Murray the younger had invested in them reluctantly and only at Somerville's insistence. This insistence may have rested on the belief that up-to-date accounts of science, especially those that presented vegetable and animal life, included large numbers of illustrations. But in some ways, these illustrations limited imagination at the same time that they added precision. They tended to abstract the organisms from any kind of recognizable context and to portray them in a static fashion. Another difficulty may have been that literary traditions associating the microscopic and telescopic worlds were more difficult to relate to subjects such as "the ear of a crab" (fig. 146), "the sucker-plate of a sea egg" (fig. 139), and "the foot of a cockle" (fig. 166), at least when these are presented to the physical eye rather than the mind's eye.

Although *On Molecular and Microscopic Science* was generally well received, it dealt with somewhat controversial subject matter in a period of rapid change. Early on, John Murray the younger, who had succeeded his father in the Murray publishing house, had reservations. His correspondence with Somerville regarding the book gives a sense of the difficulties the situation presented to young Murray and the genuine concern he had for Somerville's reputation. In December of 1868,[6] he wrote to Somerville:

> I have been alarmed by the discovery of numerous serious errors in the sheets wch [which] have already thrice before been submitted to you and wch [which] have escaped your notice. It is quite evident that it is too severe a trial of your eyesight to read over the proofs. (Somerville Collection, Folder MSBUS-3 in Dep.c.373)

---

[6] The exact date cannot be determined with certainty, but it is likely that the letter was written on December 19, 1868.

He indicates that he sought out an unidentified "competent man of Science" to do corrections, but the person he asked declined because there were many errors. It would need, in that person's opinion, "to be rewritten in great part," largely because of changes in "nomenclature of terms." He is reluctant to abandon the project because of the £200 he has already spent on illustrations but hesitant to proceed because

> your great and deserved scientific reputation is at stake – it would be a great calamity for you to publish anything wch [which] would compromise it. I would not hesitate now to proceed if I could find the way to relieve your work from error and antiquated forms of Science. (Somerville Collection, Folder MSBUS-3 in Dep.c.373)

The correspondence suggests that Murray was not able to evaluate the accuracy of the content himself. As a way of resolving the situation, Murray suggested that John Herschel review the manuscript. Murray wrote to Somerville again in January of 1869, greatly relieved because he had received a positive response to the manuscript from Herschel, who made some corrections and deletions on parts dealing with physical science. Murray also reported that he had retained a Mr. Moore to correct the parts dealing with botany and that Somerville would see the final version before it went to press. Moore and Herschel are acknowledged in the preface to the book, as are Mr. Gwyn Jeffrey, Professor Tyndall, and a Mr. Higgins, all of whom she describes as having "aided in revising some of the sheets for press." (MMS, 1: preface n.p.) In many respects, this process was similar to the one Somerville had often engaged in, in which she circulated manuscript drafts to associates who were intimately involved with the fields in which she was writing. The problem seems to be that most of her consultants, with the notable exception of Herschel, were gone, and that she had, to a great extent, become removed from the elite of the scientific community and the advantages it offered. Since young Murray handled so many of the transactions himself, it was he rather than Somerville who benefited from the exchanges with practitioners.

Although it would be tempting to interpret the negative responses that Murray received from the unnamed "man of Science" as evidence that Somerville was somehow losing her edge, the reviews suggest otherwise. On February 6, 1869, *The Athenæum*, for example, asserted that "every one will rise wiser and better from the attentive study of these volumes." ("On Molecular" 1869: 203) The problem seems not to be that Somerville is presenting inaccurate information but rather that she is not interpreting it as reviewers think she should. H. E. Roscoe, writing for the *Edinburgh Review* (1869), takes her to task for not placing "the great law of the *conservation of energy*. . . . so prominently in the foreground as might be advisable" (138), while the reviewer for *The Athenæum* believes that she has not used the proper conception of a molecule. ("On Molecular" 1869: 202) This reviewer gives her credit for having correctly identified the direction in which science is moving. "Having, by mind and method, bridged over millions of miles in space and determined some of the great phenomena of distant worlds" scientists have set out "to elucidate the mysteries of a molecule." (202) The reviewer is disappointed, however, that she does not fully come to grips with the powers involved and does not connect molecular science with microscopic science. She had successfully portrayed the sublime in apparently mundane or invisible phenomena, but she had not presented a unifying framework in which molecular and microscopic phenomena could be viewed as analogous, even though they were clearly on similar scales. Thus, she could not do what she had done in other works – she could not present the view of the world that was emerging from frontier science, in large part because the details of that view had not yet fully emerged.

Somerville was, nevertheless, using the findings of molecular and microscopic science to buttress some of her most interesting philosophical commitments – beliefs in the uniqueness of human moral and religious character, the possibility of some form of immortality for animals, and the dominance of mind over matter.

In an argument that was anticipated in *Physical Geography*, she says, "Every atom in the human frame, as well as in that of other animals, undergoes a periodic change by continual waste and renovation; but the same frame remains; the abode is changed, not the inhabitant." (MMS, 2: 11–12) She also argues that the difference between human and animal intelligence is in degree, not kind (MMS, 2: 11), and that human inability to comprehend animal communication and intelligence does not mean that they do not exist (MMS, 2: 12). She also seems to believe in a kind of immortality for animals.

> It is generally assumed that the living principle of animals is extinguished when the abode finally crumbles into dust, a tacit acknowledgment of the doctrine of materialism; for it is assuming that the high intelligence, memory, affection, fidelity, and conscience of a dog, or elephant, depend on a combination of the atoms of matter. To suppose that the vital spark is evanescent, while there is every reason to believe that the atoms of matter are imperishable, is admitting the superiority of matter over mind: an assumption altogether at variance with the result of geological sequence. (MMS, 2: 12)

She echoes Lyell, whom she interprets as presenting "us with a picture of the ever-increasing dominion of mind over matter." (MMS, 2: 12)

This use of science to buttress previous philosophical commitments and to interpret science to provide a world view was not unique to Somerville or new to *On Molecular and Microscopic Science*. Somerville had used an interpretive frame for science throughout her works, and many scientists of the nineteenth century, including those involved in thermodynamics, intertwined science with social and religious thought.[7] Somerville had used a number of traditions to legitimize science and vice versa. What is most significant about her assertion of the dominance of mind over matter is that it is a decidedly nonmaterialistic argument made in an age that was becoming increasingly materialistic. Her argument was supported by many in the 1860s but much less so in the decades following. (Zaniello 1988) Although the interpretive framework she used represented the consensus of the scientific community in the 1830s, a new consensus had emerged by the 1870s. The rhetoric of science that associated it with aesthetic, moral, and religious values had done its work and was replaced. The scene had indeed changed and the dramas of nature would be played out on a very different stage than that provided by the cosmic platform and tracing the mazes.

---

[7] See, for example, Greg Myers, "Nineteenth-Century Popularizations of Thermodynamics and the Rhetoric of Social Prophecy," *Victorian Studies* 29 (Autumn 1985): pp. 35–66.

5

Mary Somerville on Mary Somerville

Personal Recollections (1873)

No analysis is so difficult as that of one's own mind
– Mary Somerville, *Personal Recollections*

The last page of the second draft of Mary Somerville's autobiography, written near the end of her long life, contains only a few lines. These lines are recorded in an elderly hand that deteriorates rapidly as it moves down the page and seems to contradict the words themselves.

> I have every reason to be thankful that my intellect is still unimpaired, and, although my strength is weakness, my daughters support my tottering steps, and, by incessant care and help, make the infirmities of age so light to me that I am perfectly happy, *and as a memorial of my gratitude and love, I dedicate this my last work to them.*
>
> Mary Somerville

The page is very wrinkled and appears to have been crumpled and then smoothed out again. The words printed here in italics are omitted from the published version. The circumstances that led to the crumpling of the page and the omission of the final words remain, like many other things about Mary Somerville, somewhat of a mystery.

Personal Recollections holds particular interest because it sheds light on her published works and also offers answers to questions left largely unanswered by other sources. What kind of a person was Mary Somerville? What kind of life did she lead? What were her regrets and fears? How did she view her own career and abilities? What were her views on women in science and society? This chapter uses both the published and the manuscript versions of Somerville's autobiography, along with correspondence, to illuminate some of the terra incognita of Mary Somerville's character, life, and world view, all of which were considerably more complex than either her scientific writings or her personal conduct suggested. *Personal Recollections* is also interesting as an example of Victorian autobiography and as a window into science and culture in Great Britain during the nineteenth century. But it is not easily comprehended within the framework of her other books.

Even though Somerville thought of *Personal Recollections* as her "last work," its subject matter and genre are radically different than those of her other works. *Personal Recollections* contains some accounts of scientific discoveries and achievements, but it conveys and reflects very little of the substance Somerville wrote on in her other books and is not clearly recognizable as a scientific autobiography. There is much discussion of dinner parties, trips abroad, visits to friends and relatives, and where holidays were spent. These events are more important than they might first appear, given the importance of social contacts for scientists of Somerville's time. Still, *Personal Recollections* gives very little insight into the crucial turning points in the development of her scientific thought or style of presenting science.

On the other hand, it has a richness of its own in that it reveals a great deal about the woman behind the carefully constructed narrative persona and public image, both of which had acquired a somewhat mythic character by the time *Personal Recollections* was written and published. In addition, its identity as a woman's autobiography makes it possible to consider Somerville within a large and interesting genre that presented distinctive challenges for Victorian women.[1]

Another obstacle to full appreciation of Somerville's autobiography is what has been termed the "myth of personal coherence," which suggests that a biography or autobiography should be "an ordered narrative" that expresses the "essence" of the subject. (Shortland and Yeo

---

[1] Interesting works dealing with women's autobiographies include Benstock (1988), Bell and Yalom (1990), and Corbett (1992). Scholars working on women's autobiographies devote considerable attention to the blurring of gender and genre and provide a rich context for thinking about the ways that Somerville transforms expectations about gender and genre.

1996: 14) As the analysis of Somerville's life and work presented thus far demonstrates, she brought together attributes that are often seen as being contradictory or in tension with each other. Although the general satisfaction and level of success she achieved suggest that she made a stable and happy life for herself, there is no reason to believe that she managed to resolve all of the tensions or contradictions that she and her work embodied. In other words, her ability to put together a viable life and career does not mean that she achieved personal coherence. It just means that she found a way to make it work – the main implication being that we should not expect total logical consistency with regard to her statements or actions and should expect instead to encounter a fair amount of ambivalence and ambiguity.

This is especially so given the fact that her autobiography exists in several versions: two handwritten drafts by Somerville herself (designated here as M I and M II) and the final printed version edited by Mary Somerville's daughter Martha, with assistance from her close friend Frances Cobbe and her publisher, John Murray the younger. This meant that the final version was the work of many hands and reflected several points of view rather than just one.[2] The differences in the versions can also be taken to reflect, first, the difficulties that Mary Somerville faced as she attempted the analysis of her own mind and life, and, second, the difficulties that her editors faced as they attempted to create a final version that would protect and promote Somerville's memory. They all were forced to confront the difficulty of controlling historical memory.

The Character, Chronology, and Structure of Personal Recollections

Personal Recollections is, as Somerville characterized it on the title page of the first version, a "memoire," a very detailed and highly personal

---

[2] In most cases, it is impossible to tell exactly who is responsible for the changes that were made. In cases where items are deleted between M I and M II, it seems clear that Mary Somerville made those decisions herself. Given the fact that Martha Somerville took the bulk of the responsibility for the editing and editorial commentary, it is likely that she made most of the decisions after her mother's death. On the other hand, the manuscript has more than one person's handwriting indicating changes to be made, and correspondence in the Somerville Collection documents a significant role for John Murray the younger as well. Also, there are many differences between M II and the printed version that are not accounted for by notes on the manuscript itself. Because of this lack of certainty, I generally refer to actions taken by "the editors."

account, including important events and notable people known personally to the author. The reader learns many interesting inside details from it – that Charles Lyell met his wife Mary Horner at the Somervilles' home (PR 145); that Adam Sedgwick arranged "A four-poster bed . . . (a thing utterly out of our regular monastic system)" for the Somervilles when they visited Trinity College, Cambridge, in 1834 (Sedgwick to W. Somerville in PR 180); and that Pope Pius VII was a much better looking man than Gregory XVI. (PR 121–122) As these details reveal, *Personal Recollections* tells the story of Somerville's life, but she does not see herself as the only or even the most important or interesting person in the story.

Nevertheless, her readers learn how she grew up, where she lived and traveled, how she spent her time and educated her children. Most importantly, they learn about the people with whom she dined, conversed, traveled, and exchanged social calls and letters. For example, she provides the reader with a charming image of the Somervilles and other guests standing "hand-in-hand round the table" of Sir Walter Scott, leading his guests in singing "in full chorus Weel may we a'be,/Ill may we never see;/Health to the king/and the gude companie." (PR 96) Especially in the last one hundred pages, there is much to interest students of Italian political and natural history. Throughout, she registers her approval for women whose pioneering work might have left them open to criticism. These women include the sculptor Harriett Hosmer (PR 305), the actress Sarah Siddons (PR 144), the writer Joanna Baillie (PR 144), and her close friend Frances Power Cobbe, a journalist, moral philosopher, and feminist. (PR 359)

Although Somerville had been assembling her papers and correspondence for years, she probably did not begin work in earnest on her autobiography until 1867 or 1868. Her conscientiousness in collecting and organizing her papers suggests that she had a well-developed sense of the significance of her own life story. She made a collection she called "Letters of Celebrities," which included an index in addition to ten full folders, some alphabetical folders, and an autograph book. She kept notes written in the course of composing her books and made some attempt to organize them for posterity. For example, some are labeled as "fragments" or "only memoranda – may be burned at my death." The manuscript was apparently well under way on August 24, 1868, when Somerville wrote to her daughter-in-law, Agnes Greig, "I have been anxious to finish my life, as, though wonderfully well for my age, I have not much time to count upon. It is difficult to write because so many remarkable scientific and other changes have taken place in my time." (Somerville Collection, Folder MSIF-17 in Dep.c.363) Various references within the text, for example, "lately entered in my eighty-ninth year" (PR 202, 347, 352), indicate that she was still writing in 1869. The first draft must have been finished before March 16, 1869, when John Herschel wrote a letter advising her to wait until after her death to publish it. (Somerville Collection, Folder MSH-3 in Dep.c.370) Work on the second draft apparently began shortly thereafter and continued at least into 1871 and probably into 1872.[3] The handwriting markedly deteriorates from the first to the second version, becoming increasingly large and shaky toward the end of the second version.[4] There is no clear indication in the Somerville papers of the sources she worked with as she wrote. She had extensive correspondence to draw on but apparently also relied on her memory of people, dates, and events. Work on the published version moved quickly after Somerville's death on November 29, 1872. By March 13, 1873, John Murray had received the manuscript and made suggestions for editing it. By October 13, 1873, the book had gone to press, and by November 28 of the same year, 1,400 of the 1,500 copies printed had been sold.[5] The reviews were numerous and generally very positive.[6]

The book that appeared in printed form was different from the manuscripts in a number of ways but remained essentially unchanged in terms of its substance and structure. Somerville had not divided the manuscript into chapters. This was done later at the suggestion of John Murray, who also suggested adding the letters. The listing of topics that appears at the beginning of each chapter is somewhat arbitrary and sometimes inaccurate. Editorial commentary, presumably written by Martha Somerville, appears at the beginning and end of the volume and is interspersed occasionally throughout; the commentary is distinguished typographically and makes up roughly 5 percent of the total text. The text is rather loosely constructed in rough chronological order, which becomes less clear and linear toward the end of the manuscript. Very few dates or ages are ascribed to events, and some of those are incorrect or inconsistent.[7]

---

[3] Near the end of the manuscript, she mentions two eclipses that occurred in 1870 (354, 360) as well as "a brilliant display of the Aurora on the evening of the 4th February, 1871." (360) In the last paragraph of the second draft of the manuscript she left a blank for her exact age at the time of her death to be entered, i.e. "ninety- ___".

[4] Somerville complained of the shaking of her hands, especially early in the morning, in her autobiography. It was one of her only complaints in old age (M I 95).

[5] Letters from John Murray to Martha Somerville dated March 13, 1873, October 13, 1873, and November 28, 1873. (Somerville Collection, Folder MSBUS-3 in Dep.c.373)

[6] These reviews are discussed extensively in chapter 6 of this book, which deals with "Memory and Mary Somerville."

[7] Difficulty in remembering dates was a problem for Somerville from her childhood (PR 29).

One of the most interesting facets of the editing of *Personal Recollections* is that it was edited not just for style, clarity, and accuracy, but that parts of it were omitted – in some cases almost "censored" – by both Mary Somerville and the editors. Consequently, some of the most interesting questions concern what was omitted or changed in the manuscripts and why. The most significant differences were in emphasis and in the interpretation of the moral or point of Mary Somerville's story. The editors apparently felt that she exposed herself unnecessarily in some of her commentary and deleted it for those reasons. Martha Somerville and Frances Cobbe were also apparently concerned with perpetuating an image of Mary Somerville as a good woman. As subsequent discussions will demonstrate, they were not particularly successful in controlling history.

What We Learn about Mary Somerville from the Published Version

Personal Recollections is the only comprehensive source of information on Somerville's early life and path to success. All other accounts of her early years, including the one presented in this book, are derived from it, and the vast majority are decisively shaped by it, as is reflected in the remarkable similarity of accounts of Somerville's life up to the time of the publication of *Mechanism of the Heavens* (1831). Even for the period after *Mechanism* was published, her autobiography is still the most comprehensive source of information, though there is a greater variety of other sources to draw on for this later period.

Much of what we learn from *Personal Recollections* has been covered earlier in this book: her early life, self-education, and the challenges she overcame in the process; the ways that family and friends either supported or hindered her efforts; the process by which she assumed a prominent place in the world of science; how she lived her day-to-day life; and the various factors that allowed her both to meet her obligations as a wife and mother and to be a productive scientist. Somerville tends not to reflect extensively on her experiences or to search in depth for the causes or consequences of the changes and events she recounts. Nonetheless *Personal Recollections* gives us a rich view of Mary Somerville as a person – what she admired, disliked, and believed.

Some of the most interesting and ambiguous parts of *Personal Recollections* deal with Somerville's self-perception. The self-image of any prominent figure is intriguing, but Somerville's self-perception is unusually interesting because her behavior, accomplishments, and demeanor posed a puzzle to many observers and because she had such an idealized public image. Those who knew her well found her entirely praiseworthy and comprehensible, but those who knew her more superficially often found it difficult to reconcile the height of her intellectual achievements with her rather simple and ordinary style of conversing and conducting herself. The comments she makes on her own character shed some light on how these apparent contradictions arose.

From Somerville's point of view, her most important personal characteristic was probably persistence, a trait that *Personal Recollections* demonstrates as well as articulates. She also sees herself as combining timidity and independence. She felt unable to argue in conversation and was reluctant to put herself in a position that would elicit critical opinions from others, yet she put herself forward as a scientific author who helped articulate points of controversy and shape consensus. All of these ideas are captured in the following passage:

> Timidity of character, probably owing to early education, had a great influence on my daily life; for I did not assume my place in society in my younger days *and lost dignity by it* (M I 135); and in argument I was instantly silenced, although I often knew, and could have proved, that I was in the right. The only thing in which I was inflexible was in the prosecution of my studies. *I never talked of them in the family circle* (M I 135). They were perpetually interrupted, but always resumed at the first opportunity. No analysis is so difficult as that of one's own mind, but I do not think I err much in saying that perseverance is a characteristic of mine. (PR 141)

This persistence lasted well into her old age, when she reported that she was still reading "books on the higher algebra for four or five hours in the morning" and was still able to solve problems. "Sometimes I find them difficult," she tells us, "but my old obstinacy remains, for if I do not succeed to-day, I attack them again on the morrow." (PR 64)[8]

This same theme of reluctance to speak publicly about her expertise is developed in a number of other contexts. Speaking of her life before

[8] In a recent study of factors related to gender and success in science, researchers "found a marked gender difference in internal qualities in respect to persistence or perseverance. Of the women we interviewed, 45 percent named these qualities as an advantageous internal factor, compared with only 20 percent of the men. Persistence in the face of obstacles may be more important for women than for men because the typical career path of women scientists contains more obstacles and difficulties." (176) From *Who Succeeds in Science? The Gender Dimension*, by Gerhard Sonnert with the assistance of Gerald Holton (New Brunswick: Rutgers University Press, 1995). The book is based on the results of Project Access at Harvard, which involved a large-scale study of American men and women who appeared likely to do well in science.

marriage, she says that she "liked a little quiet flirtation; but I never could speak across a table, or take a leading part in conversation. This diffidence was probably owing to the secluded life I led in my early youth." (PR 364) She, says more than once that "I was not good at argument." (PR 347) In her early life, she had not spoken of scientific subjects because she feared seeming ridiculous or being criticized. After her widowhood and for the rest of her life, although she felt considerably less vulnerable to the criticism of friends and relatives, she maintained the same policy: "*It never has been my custom to talk of science in my own family circle, far less in society, and being independent I did not care for criticism.*" (M I 59) She no doubt felt strong pressure to conform to womanly standards and not appear overly "learned" or opinionated, though this does not completely explain her behavior. The consequence of her reluctance to display her knowledge in public was often a response like the one of the Reverend Sydney Smith, noted stylist and sometimes sharp-edged wit, whose reactions to Somerville were reported in a letter written by Maria Edgeworth on December 3, 1843:

> Her Introduction to all those things about Light and stars and nebulae of which I know nought[9] said Sydney, but her style clear and excellent – But as to herself I never could get anything out of her beyond what you might get from any sempstress. She avoids all depth in converse – and no pretty superficial – very amiable no doubt.... Where is she? (Edgeworth 1971: 601)

Based on what Somerville reveals about her own character, it seems likely that what Smith interpreted as lack of depth or presence was in reality a conscious decision to refuse to engage in either argument or in-depth conversation with a person with whom she likely did not feel comfortable. She probably felt she had little to gain and much to lose in such an exchange.

Her reluctance to display her expertise was also related to the qualities she looked for in others, especially the importance she placed on "varied conversation," "graces of life," unpretentiousness, and sociability. She disliked pretentiousness and criticized Lady Davy, widow of Sir Humphrey Davy, who

> like many of us had more pretension with regard to the things she could not do well than to those she really could. She was a Latin scholar ... she fancied that she spoke them [modern languages] perfectly, and was never happier than when she had people of different nations dining with

---

[9] Smith refers here to *Mechanism of the Heavens* (1831).

her, each of whom she addressed in his own language . . . [and making]
Many amusing mistakes. (PR 252)

Like many other Victorians, she devoted considerable effort to evaluating other people as companions and conversationalists. She praises
Henry Holland and William Whewell for being accomplished in
science *and* literature. On numerous occasions, she praises people for
having varied conversation. She remembers Lord Macaulay for "his
charming and brilliant conversation" (PR 223) and Sir James Macintosh
for "conversation . . . so brilliant that we forgot the time, and looking
around found that everybody had left the garden" and the gate was
locked. (PR 158)

The comments she makes in *Personal Recollections* suggest that
appearing "ordinary" was not a contradiction but a conscious decision
about style, or at least a strongly developed sense of decorum that grew
out of social conventions and also out of her own temperament,
especially as it had been formed by her early experiences. She had,
from a modern point of view, negotiated her entrance into the public
sphere by epitomizing the ideals of private sphere femininity.
(Corbett 1992: 12)

But she also seems to be overlooking a significant aspect of her own
personality. There is an inconsistency between her assertions that she
was retiring and her participation in and enjoyment of social life, as
reflected in this report in a letter of July 21, 1843, to her son Wornozow
regarding a visit to Venice: "We have now seen everything, and have
become acquainted with everybody, and met with kindness and attention beyond all description." (PR 257) This inconsistency suggests that
the reticent aspect of her personality was not always dominant and
demonstrates the very difficulty she acknowledged as inherent to the
process of analyzing one's own mind.

Another key characteristic that she emphasized in the telling of her
life story was that she had a positive and resilient nature. In more than
one circumstance, her response to an emotional blow, such as the death
of a loved one, was to throw herself energetically into a project. In an
assessment that appears near the end of the book, she says, "I have
never been of a melancholy disposition; though depressed sometimes
by circumstances, I always rallied again." (PR 348) Her conduct
throughout her life and particularly in her old age supported this
assessment and illustrated the sense of empowerment she derived
from her scientific work.

Personal Recollections also reveals that she was what she called
"eerie" and that she was fearful of being alone in the dark, particularly
during thunderstorms. "Eerie," she tells us, is "a Scotch expression for

superstitious awe" (PR 65), a quality that she shared with Sir David Brewster, "one of the greatest philosophers of the day."

> We were both born in Jedburgh, and both were influenced by the super-
> stitions of our age and country in a similar manner, for he confessed that,
> although he did not believe in ghosts, he was *eerie* when sitting up to a
> late hour in a lone house that was haunted. This is a totally different thing
> from believing in spirit-rapping, which I scorn. (PR 66)

Although eeriness resembles being superstitious, it also seems related to the responsiveness, openness, and sensibility to nature that played a prominent role in Somerville's writing. In describing her experiences in Italy, she offers this account, which gives further insight into the quality of eeriness:

> One night the moon shone so bright that we sent the carriage away and
> walked home . . . an acquaintance . . . took us to the Maglio, which is close
> by, to hear an echo. I like an echo; yet there is something so unearthly in
> the aërial voice, that it never fails to raise a superstitious chill in me, such
> as I have felt more than once as I read "Ossian" while traveling among
> our Highland hills in my early youth. In one of the grand passes of the
> Oberland, when we were in Switzerland, we were enveloped in a mist,
> through which peaks were dimly seen. We stopped to hear an echo; the
> response came clear and distinct from a great distance, and I felt as if the
> Spirit of the Mountain had spoken. (PR 325–326)

As previous discussions of Somerville's scientific writing have demonstrated, the sense of nature speaking to her seems to have stayed with her throughout her lifetime.

One important feature of her self-image was her liberal political and religious views, which are much more fully developed in *Personal Recollections* than elsewhere. This liberal outlook was one of the most important bonds she shared with William Somerville. Both of them valued religious tolerance very highly, along with what they called "liberty of conscience." These attitudes are reflected and defined in the following passage from *Personal Recollections*:

> Many of our friends had very decided and various religious opinions,
> but my husband and I never entered into controversy; we had too high
> a regard for liberty of conscience to interfere with anyone's opinions, so
> we have lived on terms of sincere friendship and love with people who
> differed essentially from us in religious views. (PR 141)

As chapter 2 demonstrated, she had grown up with family members who were quite conservative in their political views, and her liberality was in part a reaction to the excesses she was exposed to as a youth. Her responsiveness to oppression arose in part from a sense of social justice but also out of Christian compassion. Nevertheless, her liberalism must be understood in its nineteenth-century context. She favored constitutional monarchy, not Republicanism, and "although I should have been glad if the people had resisted oppression at home, when we were threatened with invasion, I would have died to prevent a Frenchman from landing on our coast." (PR 71) Although her views on non-Caucasian races were not enlightened by modern standards, she participated actively in the antislavery movement. She also actively supported the effort to liberate the Italian people and unify Italy.

Her liberality extended to the treatment of religious principles in her work. These are addressed at length by Martha Somerville in the conclusion to *Personal Recollections*, on the supposition that they have not been clearly revealed elsewhere in the autobiography or other works. While it is true that Somerville gives no indication of what church, if any, she attended in her adult life, or of how she worshipped, both *Personal Recollections* and her other works convey a great deal of information about her religious views. Mary Somerville asserts that "in all the books which I have written I have confined myself strictly and entirely to scientific subjects, although my religious opinions are very decided." (PR 141) Although to a twentieth-century reader her writing may seem permeated with religious ideas, this statement would likely have seemed accurate to a nineteenth-century reader.

Her early negative experiences in attempting to learn the catechism and sitting through extremely long church services probably initiated her lack of enthusiasm for organized religion. This attitude continued into her old age and her years in Italy, which offered her ample opportunity to comment on the inhibiting and negative effects of a priesthood. She found the Anglican church much more cold and formal than the Scottish "kirk" and ultimately very difficult to relate to. Her correspondence with Frances Power Cobbe suggests that Somerville may have been a Unitarian. She never abandoned church membership but moved toward an increasingly nondenominational position through the course of her life.

The brand of religious liberalism to which both Somervilles subscribed is clearly illustrated in this anecdote from *Personal Recollections*:

A lady who came to pay me a morning visit asked Somerville what it [a cork model of the Temple of Neptune] was; and when he told her, she said, "How dreadful it is to think that all the people who worshipped in that temple are in eternal misery, because they did not believe in our Saviour." Somerville asked, "How could they believe in Christ when He was not born till many centuries after?" I am sure she thought it was all the same. (PR 124–125)

All this said, Mary Somerville was a deeply and profoundly religious person who embraced science as a pathway to God and truth. While this outlook is conveyed in all of her writings, it is articulated most fully in *Personal Recollections*. Speaking of Charles Babbage's accomplishments in mathematics, she says,

> Nothing has afforded me so convincing a proof of the unity of the Deity as these purely mental conceptions of numerical and mathematical science which have been by slow degrees vouchsafed to man, and are still granted in these latter times by the Differential Calculus, now superseded by the Higher Algebra, all of which must have existed in that sublimely omniscient Mind from eternity. (PR 140–141)

She saw no danger to her religious beliefs from science, a fact that was emphasized by Martha Somerville's editorial commentary:

> The theories of modern science she welcomed as quite in accordance with her religious opinions. She rejected the notion of occasional interference by the Creator with His work, and believed that from the first and invariably He has acted according to a system of harmonious laws. (PR 375)

Martha Somerville continues the discussion by describing one of the only controversies with which Mary Somerville was ever associated.

> In her early life . . . the controversy raged respecting the incompatibility of the Mosaic account of Creation, the Deluge, &c., with the revelations of geology. My mother very soon accepted the modern theories, seeing in them nothing in any way hostile to true religious belief. It is singular to recall that her candid avowal of views now so common, caused her to be publicly censured by name from the pulpit of York Cathedral.

In closing this discussion, Martha Somerville reiterates her mother's combination of open-minded and strong religious faith.

> She foresaw the great modifications in opinion which further discoveries will inevitably produce; but she foresaw them without doubt or fear. Her constant prayer was for light and truth, and its full accomplishment she looked for confidently in the life beyond the grave. (PR 375)

This image of Somerville is echoed in the following comments from Maria Mitchell, who met Mary Somerville on her first European tour:

> I could not but admire Mrs. Somerville as a woman. The ascent of the steep and rugged path of science had not unfitted her for the drawing-room circle; the hours of devotion to close study have not been incompatible with the duties of a wife and mother; the mind that has turned to rigid demonstration has not thereby lost its faith in those truths which figures will not prove. "I have no doubt," said she, in speaking of the

heavenly bodies, "that in another state of existence we shall know more about these things." (Maria Mitchell in Kendall 1896: 163)

Somerville herself said, "At all events free and open discussion of all natural and moral phenomena must lead to truth at last." (PR 278) Her religious beliefs seem to have been an enabling and motivating rather than limiting factor in her pursuit of science.

Somerville's references to Darwin in *Physical Geography* and *On Molecular and Microscopic Science* reveal that she resisted Darwinism very strongly even though she believed Darwin to be a man of genius who had contributed a great deal to knowledge of the natural world and its history. *Personal Recollections* offers a clearer view of both why she admired Darwin and why she objected to Darwinism. She refers to him as a man of "great talent," "profound research," and "kindly feelings" (PR 357–358) and reveals that her primary objection to Darwin's theory has to do with "the origin of the moral sense." (PR 359) She refers approvingly to Frances Power Cobbe's "Darwinism in Morals," saying, "It is written with all the energy of her vigorous intellect as a moral philosopher, yet with a kindly tribute to Mr. Darwin's genius. I repeat no one admires Frances Cobbe more than I do." (PR 359) Cobbe's central objection is that Darwin embraces a "dependent or utilitarian morality" as opposed to an "independent and intuitive morality." (Cobbe 1972/1872: 5–6) Her problem with the former is that it arises from circumstances and thus derives from a source that does not command respect, that "is merely tentative and provisional." (Cobbe 1972/1872: 10) Somerville's affinity for evidence of the unchanging and stable nature of God and moral truth is consistent with her objection to Darwinian interpretations of human development.

Personal Recollections also reveals that Mary Somerville was a strong advocate of animal rights who had a particular attachment to birds and who campaigned against vivisection and other forms of cruelty to animals. Throughout *Personal Recollections*, she offers positive and negative observations of people's treatment of their animals. She expresses concern for the "extirpation" of falcons and "many other of our fine indigenous birds" (PR 149) and expresses horror at the Italian custom of eating "singing-birds," such as "nightingales, goldfinches, and robins" (PR 238), and the widespread practice of using birds for decorative and commercial purposes. "Many women without remorse allow the life of a pretty bird to be extinguished in order that they may deck themselves with its corpse." (PR 358) The reader learns that Somerville kept dogs as well as pet birds, some of whom sat on her shoulder as she wrote and even ate out of her mouth. (PR 16, 332) She speaks on several occasions of being "broken-hearted" over the death

of a pet bird (PR 16, 66). She was active in attempting to pass legisla-
tion that would "protect land birds" and was "grieved to find that
'The Lark which at Heaven's gate sings' is thought unworthy of man's
protection." (PR 373)

She describes her involvement in the antivivisection movement,
including her efforts to have animal protection laws passed in Italy,
"the only civilized country in Europe in which animals are not
protected." (PR 363) She objected to what she terms the "wanton
cruelties" of "Majendie and the French school of anatomy," whom
she believes to have tortured animals and proved little scientifically.
She admired Sir Charles Bell, who "made one of the greatest
physiological discoveries of the age without torturing animals." (PR
193) Although Somerville believed that animals had some form of
immortality, she thought that there were some legitimate reasons
for human beings to use and destroy animals. She is not, as a
modern reader might expect, a vegetarian. Her interest in the antivivi-
sectionist movement was something she shared with Frances Power
Cobbe, who was herself a leader in the movement. Unlike some
antivivisectionists, Somerville did not see vivisection as symbolic of
science and its values.

A number of scholars have demonstrated the ways in which links
between women and experimental animals were drawn in the late
nineteenth century.[10] Hilary Rose has pointed out that Frances Cobbe
was a major inspiration for "sections of the feminist movement [who]
made a direct link between the cruelty of men as scientists to animals
within the laboratory and the cruelty of men as husbands to women
within the home." (1994: 87) Given Somerville's close association with
Cobbe and her experiences in her first marriage, it seems reasonable to
ask whether Somerville might have made a similar link. My reading of
Somerville suggests that she was less extreme than Cobbe in her views
on vivisection and that, unlike Cobbe, Somerville did not make any
connections between antivivisectionism and feminism. The antivivi-
sectionist movement had just begun to reach its pitch at the time of
Somerville's death, and Cobbe's strongest statements on vivisection
were published after Somerville died (1879, 1889, 1904). Somerville's
affection for animals, attachment to liberal principles, and friendship with Cobbe probably provide adequate explanation for the
strength of her feeling, although the possibility that she identified in
a deeper way with experimental animals cannot be discounted
altogether.

[10] For discussion of vivisection and its relation to women and feminism see French
(1975), Lansbury (1985), and Ritvo (1987).

Personal Recollections also reveals something strongly suggested in *Physical Geography* – that Mary Somerville was what we today might call an environmentalist, especially in her emphasis on the inter-connectedness of human and other systems.[11] For example, when she returned to Scotland after many years' residence in Italy and visited Jedburgh, she said, "After many years I still thought the valley of the Jed very beautiful; I fear the pretty stream has been invaded by man-ufactories; there is a perpetual war between civilization and the beauty of nature." (PR 273) She believed that the beauty of nature has value independent of human perception: "the beauty of nature is altogether irrelative to man's admiration or appreciation" (PR 378), especially since much of it could not even be appreciated before the invention of the microscope. As chapter 2 revealed, Somerville had acquired her love of birds in childhood. She says,

> The quantity of singing birds was very great, for the farmers and gar-deners were less cruel and avaricious than they are now – though poorer. They allowed our pretty songsters to share in the bounties of providence. The shortsighted cruelty, which is too prevalent now, brings its own pun-ishment, for, owing to the reckless destruction of birds, the equilibrium of nature is disturbed, insects increase to such an extent as materially to affect every description of crop. (PR 18)

She also expresses concern about the possible extirpation of the eagle from Britain, an act which she believes "will certainly be avenged by the insects." (PR 66) Her environmentalist views seem to grow out of both a close identification with nature and an in-depth understanding of the interdependency of living and non-living, human and non-human systems.

The published version of the autobiography also reveals that Somerville was particularly fond of underdogs and others who, like her, had gone from puzzlement to celebrity. She speaks with admira-tion and affection of James Veitch and David Brewster, both of whom were from Jedburgh. She describes Veitch as

> a plough-wright, a hard-working man, but of rare genius, who taught himself mathematics and astronomy in the evenings with wonderful success, for he knew the motions of the planets, calculated eclipses and occultations, was well versed in various scientific subjects, and made excellent telescopes, of which I bought a very small one; it was the only one I ever possessed. (PR 99)

---

[11] There are a number of interesting similarities between the writing and careers of Somerville and Rachel Carson. For a discussion of Carson's work, see Raglon (1997) and Lear (1998).

When she wrote an article about comets for the *Quarterly Review* (Somerville 1835) and reported that a peasant in Hungary was the first to see a particular comet, Veitch wrote to correct her, saying (referring to himself) that "a peasant at Inchbonny was the first to see it." (PR 100) She concludes her opinion of Veitch by describing him as "a man of great mental power and acquirements who had struggled through difficulties, unaided, as I have done myself." (PR 100) She later reports that Veitch and David Brewster, who had studied astronomy with Veitch, "were as much puzzled about the meaning of the word parallax as I had been with regard to the word algebra, and only learnt what it meant when Brewster went to study for the kirk in Edinburgh." (PR 102–103)

Mary Somerville on Women in Science and Society

To understand Mary Somerville in the context of debates over women in science and society, we must begin by grasping this central point: What she accomplished through her writing on science and through her public image had much greater influence and is probably ultimately more important than any statements she made on the subject, largely because it is difficult to overestimate the esteem in which she was held and because she said very little about women. In sum, her deeds mattered more than her words, but her example mattered most of all.

This perspective on Somerville's life, work, and ultimate significance were captured well in an article entitled "A Self-Educated Woman" that appeared in the *Times Literary Supplement* on November 25, 1922, almost exactly fifty years after Somerville's death and during the period when women first began to receive university degrees in Great Britain. According to the article, "Mrs. Somerville did not agitate for women's rights. She took them; one might almost say she revealed them. She wrote no book on the higher education of women. She educated herself."[12] While this assessment is accurate, it leaves out one important fact: she did include some statements on women's rights, particularly their rights to an education, in her autobiography.

Most of Mary Somerville's statements on women appear in *Personal Recollections* and deal with the obstacles to women's education. A number of her statements on women were eliminated by the editors of her autobiography, and she apparently eliminated some herself.

---

[12] This assessment, like many of those discussed later in chapter 6, also has the effect of diminishing the value of more activist approaches to women's rights.

Women's education was clearly the issue on which she had the strongest feelings, perhaps because this was the problem she experienced most directly and from which she suffered most extensively. Once Somerville gets to the point in her narrative where she married William Somerville, she rarely mentions women's issues in connection with her own life. She mentions her friend Mary Berry who exhibited "a characteristic melancholy," which Somerville attributes to the "frivolous" character of female education that ultimately limits what women are able to do with their lives. (PR 221)

Personal Recollections also reveals that Somerville shared many attitudes typical of Victorian women of the middle class, especially with regard to money, work, and child rearing. One recurrent theme is that she never wrote in order to make money, though what we know of her circumstances suggests that the opposite was true. She says early on in *Personal Recollections* that "the idea of making money had never entered my head" (PR 59) and echoes this sentiment on several occasions.

Most of her statements on women appear in the last two chapters of *Personal Recollections*. Though a number of these were omitted (printed in italics and referenced to either M I or M II below), the omissions affected the emphasis of the manuscript rather than the substance of her view. For example, the published version retains her assertion that

> age has not abated my zeal for the emancipation of my sex from the unreasonable prejudice too prevalent in Great Britain against a literary and scientific education for women. The French are more civilized in this respect, for they have taken the lead, and have given the first example in modern times of encouragement to the high intellectual culture of the sex. (345)

She then goes on to list examples and also mentions that she

> joined in a petition to the Senate of London University, praying that degrees might be granted to women; but it was rejected. I have also frequently signed petitions to Parliament for the Female Suffrage, and have the honour now to be a member of the General Committee for Woman Suffrage in London. (346)

In addition to placing hers as the first signature on John Stuart Mill's petition to obtain suffrage for women and serving on the General Committee for Woman Suffrage in London, she also supported the American suffrage movement. The available records suggest that activists approached her because of her prestige, that she usually did not initiate involvement in activist movements, and that she was a loyal supporter once she became involved.

In the manuscript version, Somerville praised the leadership of Cambridge University for their willingness to provide women with greater access to higher education and recalled how well she had been treated when she visited the university in 1834. In the printed version, almost all of this material is omitted by the editors and replaced with their editorial commentary emphasizing the progress Somerville had observed, or in their words,

> all that has been done of late years to extend high class education to women, both classical and scientific . . . [she] hailed the establishment of the Ladies' College at Girton as a great step in the true direction, and one which could not fail to obtain most important results. (PR 347)

The replacement underscores the most consistent theme in the writing of Frances Cobbe and others with regard to Somerville: that her life and conduct refuted the "old-world theory" (PR 346) that women would lose their femininity and abandon their domestic duties if they became highly educated. This debate had been ongoing in Somerville's lifetime and was at a particularly high pitch at the time of her death. A letter from John Stuart Mill written on July 12, 1869, and included in the autobiography, praises her as

> one who has rendered such inestimable service to the cause of women by affording in her own person so high an example of their intellectual capabilities, and, finally, by giving to the protest in the great Petition of last year the weight and importance derived from the signature which headed it. (345)

The editors omit the following comments with regard to suffrage and the chances of success in getting the vote for women: "*Though of less importance than education I subscribed some years ago to a petition. . . . Success is still very doubtful.*" (M II 224)

In sum, the editors delete much of what Mary Somerville actually wrote on women and insert an editorial summary that emphasizes the goals to which they were most committed: Mary Somerville as an example of the compatibility of virtuous womanhood and higher intellectual pursuits in general, and higher education for women at the Ladies' College at Girton, Cambridge, in particular. They also eliminated nearly all negative statements, comments on other countries, and discussion of suffragists other than Mill. Frances Cobbe, herself an articulate and ardent feminist, is cast in the role of moral philosopher and animal rights advocate, not as someone who has spoken out on the "Woman Question."

Where Mary Somerville put emphasis on things yet to be accomplished and on the generally poor treatment that women encountered in many parts of the world, the editors emphasized the favorable treatment Somerville received from men of science. Although the editors do leave in a statement about women being "treated worse than their dogs" by some contemporary cultures, they omit most negative statements, including this one on the status of women in Eastern nations:

> *I ardently pray for the spread of Christianity for it is among Christian nations alone that women are considered to be the companion (?) and joint heir of salvation with man. In most Eastern nations the sex is degraded to the state of animals and in savage life a woman's home is hell. Among the Hindoos after a life of wretchedness with their tyrant husband, women are burnt alive at his death as a sacrifice to his flames (?). This atrocity was put a stop to by my cousin Samuel McPherson to his infinite honor and to an immortal reward for he never met with a terrestrial one.* (M II 223)

They also omit a passage that deals with women in the medical profession, including her friend Paulina Wright Davis who was active in the American suffrage movement.

> *Among the many distinguished Americans whom I have had the good fortune to know was Paulina W. Davis. When at Naples she came to see me with her daughter and niece. She has a gentle pleasing manner, a soft voice and must have been very pretty in her youth. For twenty years she had taken a decided part in the struggle for (?) freedom in the U.S. and had practiced as a physician.*

She believes that women are particularly well suited to the practice of one branch of medicine, obstetrics:

> *Though few women have been successful practitioners in London the Faculty need not be alarmed for there are not many women that medical practice would suit, but I think one department ought to be intirely [sic] alloted [sic] to the sex. Children have been brought into the world by women time immemorial and women have occationally [sic] died and so they do now when men officiate though aided by chloroform the most blessed (?) discovery that ever has been granted to mortals (?)* (M II 223)[13]

One of the most striking statements Somerville makes about women and science comes in the middle of *Personal Recollections*, where she talks about her response to the success of *Mechanism of the Heavens*. These remarks have been circulated by Patterson and picked up by others who have written on Somerville, but they were never published as part of *Personal Recollections*. They are particularly interesting

---

[13] For a discussion of the controversy related to the use of chloroform and its connection to the women's movement, see Poovey (1988).

because they appear to be the only remarks Somerville made about the potential of women for creative activity in science.

> *In the climax of my great success, the approbation of some of the first scientific men of the age and of the public in general I was highly gratified, but much less elated than might have been expected, for although I had recorded in a clear point of view some of the most refined and difficult analytical processes and astronomical discoveries, I was conscious that I had never made a discovery myself,* **that I had no originality. I have perseverance and intelligence but no genius, that spark from heaven is not granted to the sex, we are of the earth, earthy, whether higher powers may be alotted [sic] to us in another state of existence God knows, original genius in science is hopeless in this.***" (M I 168)[14]

It is important to note that Somerville herself deleted the parts of the passage that are in bold type above from the second draft of her autobiography. This suggests that the omitted portions of the statement did not reflect a strongly held philosophical commitment, and were part of an internal dialogue in which she explored the ultimate truth of cultural assumptions about women and creativity. In a similar vein, she also could not decide how to interpret the successes and failures in her life and in the larger effort to improve the intellectual status of women. In these cases, her natural tendency for optimism and enthusiasm for progress did battle with her sense of where the women's cause really stood. She also does not seem to be able to decide on an adequate approach for making the argument for women's education.

In the end, it seems that Mary Somerville had strong feelings about improving the lot of women but had difficulty getting outside of her culture's assumptions. She was ambivalent about making comparisons of the conditions of women and slaves, largely because she also sympathized with the antislavery movement. She tended to see the treatment of women and animals as an index of the level of civilization in a culture, and to see Christianity as the best hope of ending cruelty to both. Although she briefly mentions that "British laws are adverse to women," she offers no proposals for reforming the institution of marriage or of broadening job opportunities for women (at least none beyond the comments she makes regarding governesses). Even regarding the obstacles she herself faced in obtaining the education she so passionately wanted, she did not speculate in depth on the nature of the problem, its origins, or ramifications. She apparently had a limited conception of the forces that contributed to women's lack of accom-

---

[14] The words printed here in italics do not appear in M II. The entire quotation is omitted from the published version. If it had been included according to its placement in the manuscript, it would have appeared at the bottom of p. 176.

plishments in science. Those problems that she does focus on are the ones she experienced most directly (education) and extreme forms of cruelty.

What We Learn from the Omitted Passages

Some of the omitted passages reveal a well-developed sense of humor. The following anecdote also displays both the status Somerville achieved and the attitude with which she and her family approached her celebrity.

> A lady asked me to receive a friend of hers from the United States who wished to make my acquaintance. He was accordingly announced the next day and to my surprise entered followed by six or eight American friends, all of whom were presented to me. When he had conversed with me for a few minutes, one of his companions said you have talked with Mrs. Somerville long enough; now I want to talk with her; so down he sat; by and bye another came and said, give me your place for it is my turn to have a talk with Mrs. Somerville. The gravity with which this was carried out made it the more comical. I did not dare to look at Somerville or my girls all the time for fear I should laugh out right. That evening I told Mrs. Butler well known as Fanny Kemble that a party of her American friends who were no doubt doing Rome had come to have a talk with Mrs. Somerville. O yes, she said, they came directly from you to me and enacted the same comedy. (MI 217–218)

Many of the omitted passages deal with her fears and regrets. Probably the most significant of these is her fear that she had contributed to the death of her daughter Margaret in 1838. She says, "I felt her loss the more acutely because I feared I had strained her young mind too much. My only reason for mentioning this family affliction is to warn mothers against the fatal error I have made." (M I 151) This would have been a potentially damning statement in any era, but would have been particularly so given the nineteenth-century debate over the possible deleterious effects of mathematics and other higher intellectual activities on the health and particularly the fertility of women. Coming from Somerville, such a statement might have had a significant negative effect on women's higher education in England. It is not difficult to understand why the editors omitted it. Given its single appearance and its contrast to other statements she made, it also seems likely that it falls into the category of statements written in a moment of guilt or regret that were not permanent features of her philosophy.

It is perhaps harder to discern why other passages conveying her regrets were omitted. Referring to *Connexion of the Physical Sciences*, she says "*Notwithstanding the success of this book which has gone through ten editions, I heartily regret ever having written on popular science. The calculus was my strong point, I ought to have made a new edition of the Mechanism of the Heavens."* (M I 174) At another point, she regrets not having devoted more time to painting: *"I have a picture of a storm painted by me at the age of seventeen which is better than any thing I have painted since, so I have to repent of much misspent time."* (M I 269) These comments support others in the published version that express either a general feeling about not having spent her time productively, or the specific opinion that "In writing [*Molecular and Microscopic Science*] I made a great mistake, and repent it. Mathematics are the natural bent of my mind. If I had devoted myself exclusively to that study, I might probably have written something useful." (PR 338) She also wished she had spent more time learning languages, saying,

> *I bitterly regretted having devoted my life exclusively to Science and the dead instead of living languages. Although my husband spoke for me I felt myself an incumberance in foreign society consequently I spoke little. I should have been in high estimation among the Persians who say of their women, "To speak little is silver, not to speak at all is gold."* (M I 104)

Taken together, the omitted portions of the manuscript give greater emphasis to fears that are mentioned elsewhere in the published version, her fear of lightning, for example, of which she gave no hint in her earlier scientific descriptions. Although the published version includes an anecdote developing the idea that Somerville's mother was fearful of storms and kept the Bible "on her knee as a protection," the following is omitted:

> *I mention these circumstances because they had a lasting and distressing influence on my future life, for to this day I have never got over my fear, which I am ashamed to say has nothing to do with reason. If anyone is with me I am not afraid either night or day and can admire the magnificence of the clouds and the play of the lightning, but if alone in the night loud thunder terrifies me, though I am aware that the danger is past, even before going to bed if the sky has a stormy appearance I cannot sleep. At this time my mother and I firmly believed that the path of lightning might be changed by prayer, but I have since then learned that the Deity having once fixed the laws of the different powers of nature interferes no further.* (M I 14–15)

This passage gives insight into the ways that the rational, scientific, and cultivated side of Somerville's mind sometimes did battle with the irrational elements formed during her childhood. She also says, *"I had reason lately to be glad I was deaf for there had been violent thunderstorms during the night, I heard it not, and slept soundly, but am not so much afraid as I used to be."* (M I 105)

The editors also omit passages that reveal Somerville's foibles, weak points, and womanly traits that might be judged undesirable, along with any remarks that appeared potentially embarrassing. She tells the reader what anyone who has seen her manuscripts would already know – that she is a poor speller. She says:

> I have much trouble in writing English, especially at first, and never was aware of errors till I saw them printed in my proof sheets; especially the use of shall and will, could and should, these and more yet I was complemented on the English in some of my books by the Rev. Sidney Smith, one of the best writers of the day. As for spelling I am very bad at it even now that I am writing in extreme old age. In comparing notes with my dear old friend Joanna Baillie the poetess who agreed with me in points much more important than that of spelling, we were not a little amused to find that in writing a letter we made use of words we could spell, not those we should have used in conversation. In writing for the press I generally consulted a dictionary, sometimes when lazy I trust to the compositor for spelling, always for pointing, but my MSS are well written and easy to read, but my hand shakes badly now, especially when I first begin to write in the morning. (M I 95)

They also omit the candid assessment she offers of her own appearance, which she includes after a description of how she had become sunburned while taking observations of the sun: "*I knew it was temporary. I was still very good looking and was aware of it, and notwithstanding my love of science I liked to be admired and dressed to look well but never to look younger than my age.*" (M I 150)

The editors' omissions obscure Somerville's enthusiasm for technological advancement, which seems undiminished by her environmental concerns. Given that her general views on technology had already been widely circulated, it is somewhat strange that many specific statements were almost entirely omitted from the published version. One likely explanation is that her editors sought to distance her from the practical and economic associations of technology. Another is that many of the figures she mentioned were in one way or another controversial. In either case, the omitted discussions of technological developments provide some insight into the extent of her contacts with engineering innovators and their projects. We learn, for example, that she greatly admired the railway developer George Stevenson, both for his role in the railroads she greeted so enthusiastically and for his character and determination. She describes him in the following terms:

> No man has had such influence or produced so great a revolution in the civilized world as George Stevenson by the invention of rail ways. Self educated, of the most amiable and noble character, of unconquerable perseverance and resolution he overcame the fiercest opposition the bitterest ridicule and triumphed

> *at last on the 15ᵗʰ November 1830 when he opened the Liverpool and Manches-*
> *ter railway driving the first locomotive himself at the rate of 27 miles per hour*
> *followed by seven others all of his own construction. . . . Although every endeav-*
> *our was made for a long time to defeat Mr. Stephenson's [sic] efforts, yet no*
> *sooner was his success known, than he was requested to conduct railways not*
> *only in various parts of England but even in France and Belgium.* (M I 73)

She sees further development of the railroads as inevitable and progressive, commenting, "railway companies are now so powerful that
it would be difficult to prevent them from carrying a line through the
Queen's drawing room should it suit their convenience." (M I 74) She
concludes with this interesting remark:

> *so notwithstanding the wisdom of Solomon there's a good deal new under the*
> *sun. I wonder what will be discovered and done before the year 1968 what is left*
> *to do? Perhaps to fly, if they can carry a point of resistance aloft in a balloon the*
> *Mr. Glashier of that day may become acquainted with "the man" in the moon*
> *and his dog.* (M I 74)[15]

She toured the works of other notable engineers including Brunel,
Maudsley, Rennie, and Nasmyth and speaks admirably of their work
and heroic character. Unlike the German feminist Louise Otto and other
advocates of women's rights, she apparently saw little if any direct
connection between the advancement of technology and the liberation of women.[16]

What the Editorial Strategy Reveals

Like many other editorial decisions the three editors made, their omissions from Somerville's autobiography were apparently designed to
protect her identity as an exemplar of Victorian womanhood and as an
eminent scientist. The omissions are particularly interesting from a
modern point of view because their content seems to give a more complete view of Mary Somerville as a person and of the existential choices
she made, the ways in which she struggled for self-definition, and the
extent to which she both did and did not achieve it.

Perhaps the most significant passage the editors omitted was the
first – the introduction that Mary Somerville provided for both drafts

---

[15] These irregularities of sentence structure are in the original and are consistent with
Somerville's enthusiasm and the nature of the developments she describes.

[16] See Ingrid H. Soudek and Kathryn A. Neeley, "The Match and Other Agents of Liberation: The Role of Technology in the Social Thought of Louise Otto," *Research in Philosophy and Technology* 14 (May 1994): pp. 119–137.

of her autobiography and that seems intended as an interpretive frame-work for her life story. This introduction is very brief and quite direct in stating her motive for writing and the way in which she thinks her life should be interpreted:

> *My life has been domestic and quiet, I have no events to record that could inter-est the public, my only motive in writing it, is to show my country women that self education is possible under the most unfavorable and even discouraging cir-cumstances.* (M I and II 1)

Like many other authors of autobiographies in the Victorian period, Somerville casts herself as providing a model of self-education from which others might benefit. The editors not only significantly expand the introduction but also reframe and thus reinterpret Somerville's life story. They retain the obligatory "trope of modesty" with which numerous women's autobiographies begin, but transform Somerville's direct statement to her countrywomen into a series of more oblique statements about Somerville's own struggles and achievements. They add material establishing Somerville's aversion to revealing her private life or that of her closest friends; her beauty and simplicity of charac-ter; her religious faith; her domestic accomplishments – including an ability to produce remarkable embroidery and lace; her capacity for moral indignation; and her selflessness. They present her as the epitome of Victorian womanhood.

To understand their strategy, we must first understand their con-cerns, many of which arose from the demands of writing and the risks of publishing an autobiography. The most significant of these was the high status of the author, who was working in a genre in which she had not yet proven herself. From one point of view, it could be said that she had more to lose than to gain from publishing it. Her fame carried a certain mystique that could easily vanish once the reading public came to see her as a "real" person. What there was to gain was the opportu-nity to use Somerville's example to promote women's interests. An autobiography also presented the possibility of expanding her influence through time. As Margaret Herschel (wife of John) wrote to her on January 5, 1869, "Such as you are make the chainwork [unintelligible] which links generation to generation – or rather you overlap the links, for you spread your influence far into succeeding time." In another part of the same letter, John Herschel encouraged her to write an autobiography suggesting that the five books would "altogether form one of the most wonderful series of specimens of female authorship on record." (Somerville Collection, Folder MSH-3 in Dep.c.370)

After he had reviewed the first draft of the manuscript, Herschel's tone and approach changed. This change is reflected in a letter of

March 16, 1869, in which he conveys his assessment of the draft to
Somerville and discusses her plans for publication:

> The present generation know already how thoroughly you have proved
> it's possible to conciliate the duties of an exemplary wife, mother, and
> friend with an inner life of scientific study and deep thought – but the
> next generation and those who come after will need to be told it – or
> what is better – to gather in from just such an unostentacious [sic] self-
> revelation as the narrative part of these pages affords. (Somerville Col-
> lection, Folder MSH-3 in Dep.c.370)

He advises her to wait until after her death to publish because "the
moral and social lesson it conveys would come more emphatically, and
in some respects more advantageously in posthumous form."
(Somerville Collection, Folder MSH-3 in Dep.c.370) He explains no
further than this.

John Murray and Martha Somerville also seem to have seen some
risks in publishing the autobiography, although they resolve in the end
to go ahead. In a letter of March 13, 1873, Murray writes to Martha:

> I did not send to the press your mothers [sic] memoire until after serious
> consideration. I had come to the conclusion, that although they would
> not *add* to your mothers reputation, they contained nothing wch could
> detract. The worst that could be said against them is that they are very
> slight, in some few cases perhaps trivial, but on the other hand there is
> much to interest those to take an interest in her, in the personal traits,
> and the details of her almost self education. (Somerville Collection,
> Folder MSBUS-3 in Dep.c.370)

He suggests including letters "from her Scientific friends" to give "a
little more character to the book," along with some of Somerville's own
letters. He also says that Frances Cobbe has "a deep sympathy" with
the project and suggests that she might offer advice on it. Given his difficult experience with *Molecular and Microscopic Science*, Murray is
understandably anxious. Since Cobbe was both a highly skilled professional writer and a friend of Somerville's, she was a logical choice
to help in the effort. Her participation no doubt eased Murray's mind,
though they clearly did not agree on everything, including how much
space should be devoted to animal rights and vivisection, which were
favorite causes of Cobbe's.

One underlying theme behind many of the concerns about the
autobiography is that speaking in her own voice might damage Mary
Somerville's reputation. It was one thing for her to seem simple,
natural, even ordinary in conversation. It was apparently quite another
to seem simple, trivial, or ordinary in print. Many of these concerns, of
course, were related to the fact that she was a woman whose reputa-

tion depended as much on adhering to "the decorums of private-sphere femininity" (Corbett 1992: 94) as they did on distancing herself from the flaws believed to be common to women's writing. She had been a public figure for many years, but she had not revealed much of herself or called attention to herself as a woman author. Writing about herself presented a new set of problems.

A great deal has been written about "the anxiety of authorship" (Corbett 1992: 56) pervasive among women authors, especially in the nineteenth century, when femininity was identified so very strongly with the private sphere. To become an author was to move indisputably into the public realm, thus creating a divided identity. The typical solution to this divided identity was to "configure [feminine] authorship as congruent with the norms of domestic femininity." (57) This was the strategy Somerville's editors were pursuing when they transformed her from an advocate of women's education into the "self-less subject" of a discreet domestic narrative. It was also, to a certain extent, what Somerville herself did in a rather contradictory definition of her life as "domestic and quiet" with nothing of public interest involved in it. But one thing that Somerville did not do was to present her scientific work as an extension of her domestic role. I am not suggesting that the image the editors promoted and emphasized was manufactured; the elements were already parts of Somerville's image. The key point is their choice to emphasize Somerville as an exemplar of "true womanhood" and obscure her message of encouragement to other women who sought education.

Overall, the editors' decisions have greatest impact in the way they ultimately frame the narrative and in the last two chapters. While much is omitted – especially her thoughts on women, the progress they have made, and their prospects for the future – little is said in the omitted excerpts that was not hinted at elsewhere. As with the additions, the primary effect of the omissions is to change the emphasis. In the published version, more is said about the eruptions of Mount Vesuvius than any other single subject, and more is said about the rights of animals than about the rights of women. The omissions also modulate the tone. Nearly all negative statements, including comments on other cultures, are eliminated.

The main effect of the editorial changes was to promote a memory of Mary Somerville as a good and virtuous woman. The introduction provided by Martha Somerville provides two images of her mother that were picked up by many commentators and reviewers, in large part because they resonated with the way Somerville presented herself. One of these we have already considered – the image of a lonely child wandering the seashore and watching the heavens. The second captured

Somerville's love for nature and her profound religious beliefs, along with the connections between the two.

> All things fair were a joy to her – the flowers we brought her from our rambles, the seaweeds, the wild birds she saw, all interested and pleased her. Everything in nature spoke to her of that great God who created all things, the grand and sublimely beautiful as well as the exquisite loveliness of minute objects. Above all, in the laws which science unveils step by step, she found ever renewed motives for the love and adoration of their Author and Sustainer. This fervour of religious feeling accompanied her through life, and very early she shook off all that was dark and narrow in the creed of her first instructors for a purer and happier faith. (PR 5)

This second image also subtly emphasizes Somerville's liberalism in matters of religion.

The creation of such an idealized image is consistent with Victorian biography generally and with Somerville's status. None of the reviews I have seen question the authenticity of the virtues *Personal Recollections* attributes to Somerville. Referring to the admission of flaws in autobiographies, a writer for the *Pall Mall Gazette* on February 23, 1874, argues:

> There are no such admissions in Mrs. Somerville's "Recollections"; and the reader may therefore conclude, so evidently truthful is the narrative, that this admirable woman was free from the defects, moral and intellectual, which so often throw a shadow over the careers of illustrious persons. The beauty of Mary Somerville's life is as remarkable as the genius and energy through which she found her way to high consideration among the philosophers of her age.

She was a venerated figure to whom all three editors were tied in a number of ways and in whom they had all invested for a variety of reasons. They all were concerned with protecting an image and reputation. They did not want to include anything that would make her look foolish, vain, weak, or otherwise excessively human. After all, *Personal Recollections* was a book about a heroine, a symbol who could be permitted few regrets or doubts. It is likely that concerns like these underlay Herschel's advice about posthumous publication. In a similar vein, the editors were very much concerned with avoiding controversy, especially controversy related to the Woman Question, that might detract from Somerville's memory and prevent her from serving as a role model for the causes that mattered most to them. There is a lingering sense that they were concerned more with avoiding criticism or negative outcomes than with ensuring positive ones. Ensuring the endurance of her reputation as a scientist does not seem to have been a high priority, and may well have seemed unnecessary.

Conclusions

The Personal Recollections of Mary Somerville reveal what made Somerville good but not what made her great. They are frequently political or topical, rarely philosophical or technical, and only occasionally sublime. The book offers few glimpses into either her mathematical ability or scientific expertise. The voice she had created in her scientific writing made it somehow possible for readers to forget momentarily that she was a woman. The voice in which she spoke in *Personal Recollections* rarely made that possible. *Personal Recollections* conveys little of the substance of Mary Somerville's scientific and literary accomplishments, though it does provide evidence that she retained her intelligence, enthusiasm, and descriptive capacity to the end of her life. It was not a scientific or literary autobiography; it *was personal*. And though a discerning eye can read into it the outlines of a strategy for creating a significant public space in the masculine world of science, its effect is predominantly that of a woman "just like ourselves."

One of the most telling points emerging from an analysis of *Personal Recollections* in relation to Somerville's other writing is that Somerville handled the anxiety of authorship in a very interesting way. In *Personal Recollections*, she fully reveals the anxiety she experienced by recounting several instances in which she threatened to burn manuscripts if there was any chance that they would embarrass her. As in the case of other women authors, her first efforts were very much mediated and supported by masculine authority in the form of Henry Brougham, William Somerville, John Herschel, and Charles Babbage.[17]

This revelation of anxiety stands in sharp contrast to the narrative persona she assumed in her scientific writing. Although she made occasional, superficial gestures such as dedicating a book to the Queen or saying that she hoped it would aid her countrywomen, she never acknowledged any conflict or contradiction between her identity as a woman and her role as a scientific author. This was true in *Mechanism* where her role as translator represented a relatively small move into the public realm and it remained true as she became increasingly the

---

[17] One variation in Somerville's story is that these men gave no indication that writing was not a proper sphere for a woman, and Somerville makes no attempt to represent her writing as domestic labor (Corbett 1992: 85).

purveyor of her own formulations. This ability to manage her anxiety and keep it out of her scientific writing was clearly a major factor in her ability to establish her credibility as a scientist. As John Herschel's remarks in his review of *Mechanism* reveal, critics were accustomed to hearing expressions of anxiety from female authors, though they seemed entirely unaware that this anxiety was culturally induced and virtually required of women. Somerville's writing surprised them by revealing none of that anxiety.

Somerville's autobiography tended to mark a transition in which she moved from "great" to "good" in the public eye. The anxiety of authorship that she revealed in *Personal Recollections*, but that she appeared not to have experienced to any great extent while writing it, may have contributed to that shift, as did the interpretive frame in which the editors located her life.[18]

---

[18] One of the interesting features of the publication history of *Personal Recollections* is that, although Somerville still seeks advice about publishing it, she seems by then to have a strong sense of her own authority and less anxiety than her editors.

6

Memory and Mary Somerville

In the Public Eye and Historical Memory

Mrs. Somerville's reputation is likely to be permanent, but it is possible that this unaffected record of a beautiful and consistent life may be of more benefit to society than even the valuable works to which she is indebted for her fame.
— Review of *Personal Recollections*, *The Spectator*, January 31, 1874

I began this project with the naïve assumption that Mary Somerville was a forgotten woman. My reasons for thinking she was forgotten were straightforward: I had read dozens of comprehensive histories of science, and I had never seen her treated as anything more than a footnote, and rarely even as a footnote. As I examined in detail how she had been treated by the history of science, I realized that something much more interesting than forgetting was going on. She was not forgotten. She just could not be integrated into the patterns of traditional history of science. The eminence she had achieved in her own time was overshadowed by the fact that there was no major discovery to which her name could be attached and by the conceptual filters that either discounted women's achievements or failed altogether to recognize them.

The "Forgetting" of Mary Somerville

Of all the aspects of Mary Somerville's life and career, her fall into relative obscurity – what I will call here her "forgetting" – is one of the most interesting. By "forgetting" I mean not the literal loss of knowledge of her existence but rather her absence from standard histories of science. During her lifetime, she had been an integral part of the living scientific community, but she was an outsider to history.

As the remarks from *The Spectator* with which this chapter opens attest, it would have been hard for Somerville or her contemporaries to imagine that she could be forgotten. The numerous awards she received near the end of her very long life, the effort she invested in her autobiography and the collection of her papers, and the often reverential attitude with which she was approached all suggested the probable permanence of her reputation. But *The Spectator's* remarks also reflect a tension between two ways of seeing Mary Somerville's life. Was it best understood as "a beautiful and consistent life" or in terms of "the valuable works" she produced? If the decision were in favor of her works, could her case be used to support generalizations about women and their intellectual ability? These questions were particularly prominent in the debates of the time.

The way Mary Somerville's life was interpreted was significantly affected by a number of processes already underway during her lifetime, many of which intensified in the years following her death. Perhaps the most significant of these was the recognition that the debates about women's education were really debates about the role and power of women in society. Education was increasingly recognized as the "thin end of the wedge" by which women would gain greater power. (Blacker 1996: xiii)[1] As women's opposition to the social restrictions they faced became more coherent and articulate, it was countered by stronger and more vehement resistance. This increasing contentiousness manifested itself as resistance to allowing Mary Somerville to be treated either as an exemplar of what a feminine intellect could achieve if given the opportunity or as a symbol for the late nineteenth-century women's movement.

---

[1] Carmen Blacker's preface to *Cambridge Women* (1996) provides a vivid account of the debates about women intellectuals as they played out in the history of Cambridge University.

But there were other factors involved as well. As science began to gain significant cultural influence, the tendency to define it in exclusively masculine terms was intensified. Moreover, specialization became increasingly synonymous with the serious pursuit of science. The heroic discoverer model of the history of science was already in place during Somerville's lifetime. By the end of the century, it would become the dominant model for narrating the history of science. These developments combined to create a situation in which it was very difficult for Somerville's achievement to be fully recognized and appreciated.

Added to these trends was the secularization of intellectual life generally and of science in particular. The reader cannot get far in Somerville's writing without encountering the notion of a scientist as someone who, in the words of Pope's *Essay on Man*, "looks through Nature up to Nature's God" (line 331). During the debates that followed the publication of Darwin's *Origin of Species* in 1859, the notion of the compatibility between science and religion was challenged and eventually so radically transformed that religious issues and themes came to be viewed as "tainting" science and scientific writing. (Zaniello 1988) Somerville's perception of God in the processes of the universe was an additional pleasure and motivation for her pursuit of science, not its primary aim. Nevertheless, the religious overtones of Somerville's writing and her refusal to accept some of the implications of Darwinism meant that her work came to seem old-fashioned very quickly.

The "Brightest Star" in the Intellectual Constellation

Mary Somerville had been a prominent public figure for more than forty years at the time of her death. The obituaries that appeared after her death in 1872 and the reviews of *Personal Recollections*, which was published late in 1873, provide ample evidence of the very high esteem in which Somerville was held.[2] On December 2, 1872, the *Morning Post* emphasized the longevity of Somerville's career and reputation: "Long before many of the distinguished cultivators of science now among us were born, Mrs. Somerville had already taken her place among the

---

[2] For a discussion of the significance of obituary notices in providing insight into the ways a biographical subject is perceived by contemporaries and successors, see Geoffrey Cantor, "The Scientist as Hero: Public Images of Michael Faraday" in *Telling Lives in Science: Essays on Scientific Biography*, ed. Michael Shortland and Richard Yeo (Cambridge: Cambridge University Press, 1996).

original investigators of nature." Richard Owen's letter to Martha Somerville, written on December 6, 1873, just after he read *Personal Recollections*, conveys the way she was viewed by her fellow scientists.

> I sat up late . . . living over again a large part of my working life, in the reminiscences of all my honoured fellow-labourers, or almost all, both at home and abroad, through the fellowship with them of the brightest star in that Intellectual Constellation – brightest through the beauty of mind, of nature, of person, which were associated and concentrated in your dear Mother. (Somerville Collection, Folder MSO-1 in Dep.c.371)

Her position as "the brightest star in that Intellectual Constellation" – and quite a constellation it was – depended on the combination of characteristics that Somerville brought together rather than on a single trait or accomplishment.

In their assessment of Mary Somerville's life and accomplishments, the reviewers and writers of obituaries were unanimous on one very important point: that she had demonstrated the compatibility of high intellectual attainments in mathematics and science with feminine ideals. The *Morning Post* of December 2, 1872, said she had proved "how much a woman could and can accomplish in intellectual pursuits without sacrificing of her usefulness in the sphere of home or her feminine dignity." On January 4, 1874, the *New York Herald* described Somerville's life as "an instance proved of a woman renowned for intellectual gifts and accomplishments, achieving and creating happiness as a mother." A writer for *Nature* in a review published on April 2, 1874, characterized *Personal Recollections* "as the record of a life in which the fulfillment of all the natural and conventional claims upon a woman's time was combined with practical and theoretical pre-eminence in the most abstruse departments of physical inquiry."

This ability to demonstrate the compatibility of traditional womanhood with the pursuit of science was, as the preceding chapters demonstrate, a crucial factor in Somerville's rise to prominence. Traditional womanhood was a symbol of stability, and the association of science with stability and tradition was particularly important in an era when almost everything seemed to be changing – largely as a result of science. Somerville's character and accomplishments served during the early Victorian era as examples that could be used to sell the idea of science as a cultural resource. As a number of scholars have demonstrated, "The social legitimation of the nascent scientific enterprise, and its bid for increased cultural status in the nineteenth century, depended heavily on claims about the kind of people it attracted." (Shortland and Yeo 1996: 37) Somerville's capacity to function as a symbol played a large role in the acceptance Somerville gained from the scientific elite.

Through the course of Somerville's lifetime, gender roles became increasingly important for organizing and stabilizing social life. If she had challenged traditional feminine ideals, I believe she would never have achieved the stature she did. But there was a drawback to her acceptance of these ideals. As the reviews and obituary notices demonstrate, her conformity eventually became part of a powerful argument designed to exclude women from science.

A Story with a Multiplicity of Interpretations

Although Somerville's combination of intellectual pursuits and feminine behavior was the central theme in the reviews and obituaries, there were other common themes. Despite the commonalities in the way they described Somerville and told her story, these writers saw that story as supporting a range of conclusions. The common themes they stressed and the multiple interpretations they put upon Somerville's life both reflected the factors that made Somerville great and foreshadowed the forgetting of Mary Somerville. This multiplicity of interpretations arose from the strategies that Somerville used to present herself to the public and from the wide range of important issues to which her life could be related.

Her conduct with regard to the June 1832 meeting of the British Association for the Advancement of Science (BAAS) provides a typical example of her approach and the ambiguity it created. The recent publication of *Mechanism* had established Somerville as one of the scientific elite in Britain, and her absence from the meeting would have been noticed.[3] On the other hand, there was great concern that the presence of women "would turn the thing into a sort of . . . dillettanti meeting instead of a serious philosophical union of working men." (Buckland in Morrell and Thackray 1981: 150) There was no suggestion that Mrs. Somerville was a dilettante, but there was clearly an underlying belief that welcoming one woman at the meeting would signal that all women were welcome, thus feminizing the BAAS and undermining its status.

In the end, Somerville did not attend the meeting. She also never gave an explanation for her absence. John Herschel also stayed away from the 1832 meeting, and it is likely that his behavior influenced hers. She also may have been reluctant to make the organizers of the meeting uncomfortable, since she counted many of them among her friends. Attending the meeting would have been a rather public and

---

[3] Sotheby's poem (1834) supports the contention that she would have been missed.

confrontational act, and Somerville tended to avoid public confrontation, preferring instead quiet persistence. On the one hand, she had taken her daughters to the lectures in geology by Charles Lyell that had recently, though controversially, been opened to women at King's College in the new University of London. (This action was reversed by the Council of the college before the series of lectures ended in June 1832.) On the other hand, she had judiciously avoided taking positions or exhibiting behavior that would have been labeled "bluestocking." To a great extent, the esteem in which she was held depended on her differentiation from the bluestockings, and she must have known, at least subconsciously, that attending the meeting might put that aspect of her reputation at risk.[4]

The available evidence, including the Somerville correspondence, suggests that she stayed away from the meeting because she was exhausted and because she could not afford the expense that would be involved in keeping up appearances in accord with the position she enjoyed among the scientific elite. She confided to a friend who was also trying to manage a household with limited financial resources that the family would be "stationary all summer moving is so expensive." (In Patterson 1983: 94) To Buckland she said nothing. William Somerville apparently conveyed to Buckland his fear that his wife would not attend the meeting. In a very interesting choice of words, Buckland later reported that William Somerville "led me to infer"[5] that Mrs. Somerville believed the presence of women would undermine the seriousness of the reading of the papers. Buckland's ultimate interpretation, as reported by his daughter, was that "in the end Mrs. Somerville decided not to attend the meeting for fear that her presence should encourage less capable representatives of her sex to be present." (In Patterson 1983: 94)

I agree with Elizabeth Patterson's assessment that "nothing in Mary Somerville's character and none of her other actions lend validity to the statement of Buckland's daughter." (Patterson 1983: 95) Through-

---

[4] Examining what Somerville was able to resist without being labeled a "bluestocking" provides an interesting example of what Poovey calls the uneven deployment of gendered concepts. For an example of a late-nineteenth-century assessment of Somerville, see Bunbury, 1881.

[5] This is the wording used by Gordon in the *Life of William Buckland*. In their transcription of the letter, Morrell and Thackray use the words "informed me that." Morrell and Thackray's account of the subsequent history reveals that, although "dealing with the ladies" was a major issue for the BAAS, it was not very long until women made significant headway in the organization. In 1835, they began to disregard rules against their attending sectional meetings, and by 1840 they had access to all sections, although they still could not be members. (Morrell and Thackray 1981)

out her life, Mary Somerville showed courage in challenging assumptions about what women could do, especially in the intellectual realm. In particular, there is nothing to substantiate the kind of condescension that the statement implies. And while it is true that "Mary Somerville always supported in word and often in deed the struggle for wider opportunities for women" (Patterson 1983: 93–94), I also believe that Somerville may not have seen her appearance at the meeting as being particularly likely to advance the cause of women or as being related to that cause in any significant way. In any case, the absence of any explicit statement on her part left room for the inference Buckland later made in a letter to Murchison: "Mrs. Somerville's opinion as confirmed by her action is clearly in the negative." (Quoted in Morrell and Thackray 1981: 150) Her silence on the subject of women at the meeting left room for the observers to interpret her behavior in ways that suited their own agendas.

This strategy is strikingly similar to that used by Kathleen Lonsdale, one of the first two women elected as fellows of the Royal Society in 1945, and recounted by Hilary Rose (1994) in her discussion of women and the Royal Society. The similarities reflect the high degree of tact required in women who wish to establish themselves in male-dominated institutions. Although Lonsdale was a militant pacifist, she took a much less assertive stance with regard to the Royal Society. She indicated when approached that she was willing to be considered for fellowship in the society, but "not if it means dissension among the Fellowship." (Hodgkin quoted in Rose 1994: 126) She established her desirability as a fellow by accepting the members' authority to decide whom they would willingly admit. Rose continues, "Such personal self-effacement combined with public scientific excellence provided Lonsdale with a means for managing her gender identity which was less likely to give rise to . . . antagonisms." (126)[6]

My interpretation of Somerville's strategy for managing her gender identity is that she made the best out of a bad situation and let ambiguity work to her advantage. She let others think what they pleased and avoided embarrassment by saying nothing herself. The ambiguity she created also resolved the conflict her gender presented for the gentlemen of the BAAS. Given her circumstances, it may have been the only reasonable stance to take, but this kind of behavior ultimately left her life open to interpretations that were contrary to the interests of women and quite inconsistent with those she herself would likely have

---

[6] Lonsdale, like Somerville, had chosen her spouse wisely. Lonsdale said, "For a woman and especially a married woman with children, to become an exceptional scientist, she must first choose or have chosen a good husband." (In Rose 1994: 131).

made. Others of the interpretations, while she would likely have been quite pleased with them, played a significant role in her forgetting. These interpretations tended to fall into three categories: (1) Mary Somerville as the very type and model of womanly excellence, (2) Mary Somerville as proof of the power of the female intellect, and (3) Mary Somerville as unique. In developing all of these interpretations, the writers of the reviews and obituaries provide an extensive inventory of gender ideologies at work in the early 1870s.

Mary Somerville as "The Very Type and Model of Womanly Excellence"

This interpretation emphasized Somerville's virtue and deeply held religious beliefs. In its choice of allusions and images, *The Woman's Journal* captured the religious themes and perspective that had permeated Somerville's writings on science and that Martha Somerville had emphasized in her concluding remarks in *Personal Recollections*. Using an interesting mix of poetic, literary, and biblical illustrations, *The Woman's Journal* asserted, "The 'Pilgrim' has reached the 'Land of Beulah, where there is no more night.' Nature has led her most faithful follower 'up to Nature's God.'" (*The Woman's Journal*, December 21, 1872) Calling Somerville "the greatest mathematician among women," *The Spectator* reported that "Mrs. Somerville with all her great scientific attainments, was no sceptic [sic]." (December 7, 1872) In an obituary notice appearing in the *Echo*, Frances Power Cobbe presented Somerville

> as the very type and model of womanly excellence. She has been good and done good. Her very name has given courage to thousands of her sex. . . . She has shown to woman how good and wise it is possible for woman to be, and how well all the relations of life may be fulfilled from childhood to the last verge of mortality. (December 3, 1872)

Probably the most extravagant statement of this interpretation appeared in the *Kelso Mail* in its review of *Personal Recollections*.

> Mary Somerville, the fair, the gentle, and the good, whose sympathies extended to the meanest of the animal creation, and to the humblest flower which blossoms by the wayside; Mary Somerville, the noble, the learned, and the wise, whose unwearied research made science unfold its wonders and who saw in the works of nature not only the power but also the goodness of God. (March 25, 1874)

Even the writers who emphasized other themes devoted some space to establishing that Somerville was the exemplar of ideal womanhood.

Mary Somerville as Proof of the Power of Female Intellect

A relatively small number of obituaries and reviews argued that Somerville's accomplishments proved the capability of women for higher intellectual pursuits, though the remarks made after her death emphasized this less than it had been emphasized during her lifetime. The *Morning Post* characterized Somerville as

> a lady who has shown that it is not necessary for woman always to rest contented with superficial attainments and a subordinate place in the intelligent world, but has done honour to the female intellect by her study of nature and nature's operations. (December 2, 1872)

An article entitled "Our Obituary Record" and published on December 14, 1872, put the case even more strongly:

> She possessed the most extraordinary intellectual capacity, and her life and works constitute an unanswerable refutation of the fallacy that a woman's brain is of necessity unfitted for abstract studies; that it is impossible for a woman to be successful in scientific research, or to attain proficiency in mathematics. When, too, the believers in the equality of the sexes are twitted with the assumed fact that the pursuit of such studies invariably deprives a woman of sympathetic tenderness and unobtrusive modesty, which are her most attractive charms, they can point to Mrs. Somerville as one who succeeded in reaching the very highest distinction in the scientific world without sacrificing one iota of her feminine and household gracefullness [sic] and dignity. (Somerville Collection, Folder MSDIP-19 in Dep.c.375)

But very few interpretations of Mary Somerville's life were this bold in their assertions about women's capacity for higher intellectual pursuits. *The Athenaeum* more modestly asserted that her life proved the assertion "that women are capable of taking a higher place." (December 13, 1873)

In many cases, these acknowledgments of her intelligence are followed by assertions that the real source of female underachievement is not the absence of ability but rather lack of will and determination. *The Athenaeum* took this line when it concluded,

> Her story teaches us that difficulties ... are powerless if the mind is honestly bent on improvement; that the performance of the humblest

domestic duties is not incompatible with the cultivation of the highest
intellectual adornments; that the true path to power for women is
not to be found in the rough road of popularity, but in devotion to the
duties of the domestic home, and an unobtrusive, but persistent search-
ing after knowledge. (1873: 762)

Chambers's Journal took a similar position and expressed this conclusion
even more plainly at the end of their review of *Personal Recollections.*

The moral that may be drawn from the life of Mrs. Somerville ... is
exceedingly obvious. Considering how meagre was her education, what
chilling difficulties she had to encounter in her persevering efforts at self-
culture while never neglecting ordinary duties, and looking to the literary
and scientific eminence which she attained, we are entitled to point out to
almost the humblest of her sex, that where there is a resolute determination to improve the intellectual and moral faculties, all obstacles have a fair
chance of being successfully overcome. It is usually the *will*, more than
opportunity or natural capacity that is deficient. (1874: 36)

This particular assessment reflects the notion of the self-made quality
of genius, what Corbett (1992) has called "the mythologizing of the
individual as arbiter of his own destiny," (19) a strategy that obscures
the social and other factors that contribute to success or failure. From
this interpretation, it was only a short distance to the next most
common interpretation, that Mary Somerville was the exception who
proved the rule of women's general unfitness for demanding intellectual pursuits.

Mary Somerville as Unique:
The Exception that Proves the Rule

The writer who reviewed *Personal Recollections* for *Nature* started by
characterizing Somerville as "an illustrious woman ... a woman
unique, or almost unique from one point of view ... the one woman
of her time, and perhaps of all times" (April 2, 1874: 417) who earned
the respect of eminent men of science. Although the writer admitted
the possibility that the story of Somerville's life could "with perhaps
equal justice, be made to yield arguments for and against the claims
advanced for women's equality to men in intellectual capacity," that
possibility remained open only for a moment, and the reader quickly
becomes aware of where the argument about uniqueness is going.
Somerville is the exception who proves the rule: "We can scarcely
expect to meet with many Mary Somervilles. Her genius was unique

of its kind, and wholly exceptional, and this fact seems to have been frankly and generously admitted by all who came into contact with her." (417)

Noting that "the name of Mary Somerville has always been a tower of strength to the promoters of woman's emancipation from the enactments established by man for her exclusion," the review employs a strategy by which Mary Somerville is built up, perhaps even unrealistically, in order to diminish other women. The rhetoric of the review puts the laudatory quotations with which this book began in an entirely new light.

This strategy emphasizes how thoroughly feminine Somerville was and hints that there was something vaguely deceptive about the way she chose not to display her learning. As the writer for *Nature* described her,

> So successfully did she conceal her learning under a delicate feminine exterior, a shy manner, and the practical qualities of an efficient mistress of a household, coupled with the graceful, artistic accomplishments of an elegant woman of the world, that ordinary visitors, who sought her as a prodigy, came away disappointed that she looked and behaved like any other materfamilias, and talked just like other people. (417)

It also emphasizes the "sympathy and hearty recognition" she received from eminent men of science and the "degree in which Mrs. Somerville's acquirements differed from those of women generally at that period," all of which makes the case for her status as an exception and leads to a commonly asked question about women in science, a question that assumes that ability is the determining factor in scientific success.

> It must be admitted that it is precisely through this exceptional character of her attainments that her case may be adduced in proof of the rule that women are not by nature adapted for studies which involve the higher processes of induction and analysis. If such powers as hers had been more generally granted to women, why is she the only woman on record amongst us who has exhibited them? (418)

The review closes with a conclusion that is undermined by its own logic but probably seemed persuasive to many nonetheless:

> Mary Somerville will always present a noble instance of what *a* [emphasis added] woman has been capable of achieving, but it would be straining the argument too far to say that we are justified from her special case to draw general conclusions in regard to women's aptitude for the study of the higher forms of physical science. (418)

Of course, the writer has been drawing general conclusions on that subject throughout the review. The argument is set up to create a dichotomy: the story of her life must either make the case for female equality in intellectual capacity or refute it. The test is to be the numbers of women who have contributed. The existence of even one woman who has received a sympathetic response is taken as proof that men are willing to recognize women's accomplishments. Her uniqueness is interpreted so that it supports the overall superiority of men.[7]

The writer for *The Saturday Review*, like many others, took Somerville's nonconfrontational behavior as an indication that she shared the opinion of women's inferiority to men in intellectual pursuits. Also, as in many other reviews, this one conveys a sense of woman as adversary that had been absent from earlier discussions of Somerville. I quote from this review at length because the argument it presents constitutes an important and ultimately very powerful rhetorical strategy. Referring to *Personal Recollections*, *The Saturday Review* said:

> It is the unobtrusive record of what can be done by the steady culture of good natural powers, and the pursuit of a high standard of excellence, in order to win for a woman a distinguished place in the sphere habitually reserved to men, without parting with any of those characteristics of mind or character or demeanour which have ever been taken to form the grace and glory of womanhood. Far from setting up for herself a conscious and deliberate rivalry with men in a field of intellectual labour which is not as a rule open to her sex, or vaunting herself on the attainment of an eminence unknown to any woman of our day, Mrs. Somerville was content to give quiet scope to the tastes and inclinations which led her to the study of science. The last person to see in herself the genius who was to assert woman's desecrated rights, and win back from men their usurped dominion of the realm of knowledge, she lent no countenance to those of her sisterhood who shriek against the conventional relations which are supposed to oppress them, and who bring not a tithe of her mental power to the struggle for intellectual supremacy. It was no mere timidity of disposition, but instinctive delicacy of soul, which made her feel that there was a line drawn by nature between the spheres of usefulness or duty of women and of men. Hence the thoroughly femi-

---

[7] It is important to note that *Nature* was not uniform in its opposition to women in science. In the period shortly after its founding, *Nature* showed considerable interest in and support for science education for women. There seems to have been variety of opinion among the various writers for *Nature* where women were concerned. There also seems to have been a distinction made between science education that would prepare women for their domestic duties (this was the type usually advocated by the writers) and science education intended to prepare them for the assumption of professional status within science that Somerville attained.

nine tone which showed itself in all her tastes, and tempered her severer studies. Not a tinge of what is vulgarly known as blueness was to be detected in her demeanour or her literary work. (1874: 53)

Both the writer of the review, and, most likely Somerville herself, were unaware that her strategy of not setting up "a deliberate rivalry with men" was neither an unconstrained choice or an expression of her "nature." It was the only viable option available to her.

One implication of this argument was that if Mrs. Somerville was not complaining, lesser women had no right to complain either. If a woman with Somerville's "good natural powers" accepted the system as it existed, those who possessed "not a tithe" of her ability had no moral grounds to protest. This strategy demeans those who assert women's rights by portraying them as vulgar, shrieking women of weak intellect who fail to obey the dictates of nature and instinct. Somerville, in contrast, is portrayed as delicate, quiet, tasteful, temperate – and nonconfrontational. (The editorial strategy that Frances Cobbe and Martha Somerville followed in editing *Personal Recollections* seems explicitly designed to avoid the censure of this kind of critic.)

The effect is to raise the standard and at the same time reconfigure the discussion in adversarial terms as a "struggle for intellectual superiority." The same reviewer suggests that Somerville is the best example women have to offer to make their case for equality, but that she still is not as good as the best men in the field of science.

All that she has given to the world serves thus far to confirm the belief that, whatever may be said of the equality of the sexes, no woman's work in any department of knowledge has equalled [sic] the highest of man's work. Receptive and reproductive rather than creative, the intellect of woman, as here shown in the case of the most intellectual woman of our age, has for its function to assimilate and to give clear and easy expression to what has sprung from a brain of more robust and original power. (1874: 55)

This rhetoric, with its stereotype of the female intellect as "receptive and reproductive rather than creative," sets up an ideal of feminine behavior that is clearly non-competitive at the same time that it portrays the situation as a battle. It also figures Somerville's scientific writing as the reception, reproductive assimilation, and expression of ideas formulated by robust and original male brains, thus undermining claims about the lasting significance of her scientific work.

The language and themes of the review in the *Atlantic Monthly* are even less sympathetic to Somerville and to women generally. The author follows the same strategy of praising Somerville in order to question the potential of other women. The author emphasizes the "imposing" nature of her work and juxtaposes it with the "sweetness" of her character, and then goes on to make two arguments regarding "her extreme rarity as a phenomenon" ("Recent Literature" 1874: 494) and the willingness of men to recognize the achievements of women who really deserve recognition. Both echo contemporary debates about women's access to higher education. The conclusion is that "a woman like Mrs. Somerville is born, not made, and one is strongly impressed by the belief that no possible system of education, 'higher' or other, could ever produce her like." (494–495) The implication that follows is that the development of female intellectual ability is a spontaneous natural process on which social conditions can have relatively little effect, though the author does acknowledge that "gentle breeding . . . combined with great simplicity of life and manners" might be conducive to feminine achievement. (495)

> Where a woman feels within her a strong special aptitude irresistibly impelling her to an unusual career, she has only to strive patiently until she has achieved something positively valuable, and she may be sure of prompt and generous recognition by men. It is the unauthenticated claims, the restless demand for wider opportunities when the narrower have not been employed that provoke to resistance and refusal. (495)

Such an argument seems deliberately designed to undercut women's demand for access to higher education. In this analysis, there seem to be two kinds of women: those who give men in positions of authority very good reason to reject their claims for opportunity and recognition, and those who earn recognition without being given any special kinds of assistance. As an example of a woman who earned but never demanded recognition, Mary Somerville becomes a weapon in the rhetorical arsenal used to deny women recognition. Somerville is presented as the woman who worked hard enough to earn admission to the inner sanctum of science.

> The most precious and carefully guarded libraries were thrown open to her, and scores of learned societies, which had never admitted a woman before, elected her by acclamation to their membership. In short, all that for which too many women nowadays are content to wit and whine, or fitfully and carelessly struggle, came naturally and quietly to Mrs. Somerville. And the reason was that she never asked for anything until she had earned it. Or, rather, she never asked at all, but was content to earn. (495)

Her accomplishments and opportunities hardly came "naturally" though it could be argued that they came quietly enough. She asked deferentially, cautiously, for opportunity rather than recognition, but it is hardly true that she never asked for anything at all.

One of the most significant implications of such a view is that activism is the last resort of failures, and that the women who are really deserving will get the recognition they deserve. Thus, the assessment of Mary Somerville's life became in many cases an occasion for celebrating her uniqueness and accomplishments while at the same time making it clear that admitting Mrs. Somerville to the fellowship of science was not to be understood as an invitation to other women. The grounds of the argument often seemed to shift from the capability of women to the will and character of women and from science as a demanding intellectual pursuit to science as a seat of power.

The arguments made in these reviews demonstrate that significant but subtle changes in the rhetoric relating to science and women were taking place at the time Somerville's life and achievements were being committed to memory. Some of these same changes were played out as the possibility of Somerville's being buried in Westminster Abbey was explored. The events surrounding that request were reported by Frances Power Cobbe in her autobiography.

> Mrs. Somerville ought to have been buried in Westminster Abbey. When I saw her death announced on the posters of the newspapers in the streets of London, I hurried as soon as I could recover myself, to ask Dean Stanley to arrange for her interment in the Abbey. The Dean consented freely and with hearty approval to my proposition, and Mrs. Somerville's nephew, Sir William Fairfax, promised at once to defray all expenses. There was only one thing further needed, and that was the usual formal request from some public body or official persons to the Dean and Chapter of Westminster. (1894: 350)

But, Cobbe reports, problems arose when the Dean sought the official approval of the scientific establishment, some of whom appeared to have come to view Somerville more as a rival than an ally.

> Dean Stanley had immediately written to the Astronomer Royal to suggest that he and the President of the Royal Society, as the representatives of the sciences with which Mrs. Somerville's fame was connected, should address to him the demand which would authorize his proceeding with the matter. But that gentleman *refused* to do it – on the ground that *he* had never read Mrs. Somerville's books! Whether he had read one in which she took the opposite side from his in the sharp and angry Adams-Le Verrier controversy, it is not for me to say. Anyway, jealousy, either scientific or masculine, declined to admit Mary Somerville's claims to a place in the national Valhalla, wherein so many men neither intellectually nor morally her equals have been welcomed. (1894: 350–351)

The earlier rhetoric, including the opposition faced by Somerville as a girl and a young woman, expressed reservations about what women would be able to accomplish in science and emphasized the top priority of women developing "womanly" abilities of which science was not a part. The later rhetoric seems to assume a more adversarial relationship between women and science (and by implication women and men). Science has become a source of power and authority, and the demands of women to be admitted are a threat to that authority. Many interpreters of Somerville's life and work showed a determination to interpret her in a way that would make it clear that her success and recognition should not be viewed as a precedent for other women to follow. Her example was interpreted in nonrevolutionary terms. As long as women were limited to strictly feminine behavior and as long as the failure of women to excel in science was seen as a failure of will, none of the social and structural factors that limited women's participation in science would be addressed.

The Mary Somerville Strategy

As early as 1876, just a few years after Somerville's death, Maria Mitchell offered what now seems the most persuasive answer to the question of why so few women have made major discoveries in science, an answer that applies well to Somerville's case.

> The laws of nature are not discovered by accident; theories do not come by chance, even to the greatest minds; they are not born of the hurry and worry of daily toil; they are diligently sought, they are patiently waited for, they are received with cautious reserve, they are accepted with reverence and awe. And until able women have given their lives to investigation, it is idle to discuss their capacity for original work. (Mitchell in Kohlstedt 1987: 139)

It took roughly a hundred years of discussing women's capacity for original work for people to begin to grasp the explanatory power of Mitchell's statement. In the meantime, the examples of the few celebrated women in science were used in ways that raised the standards for all women to yet higher levels. Margaret Rossiter has described this as the Marie Curie strategy. (1982: 130) Those who used it insisted that they were open to hiring and promoting women – as long as those women were as capable or at least as promising as Marie Curie. As Rossiter explains, "before long most professors and department chairmen were interpreting Curie's example" not as supporting the capability of any woman to follow in Curie's footsteps, but to assert

> that *every* female aspirant for a faculty position *must* be a budding Marie Curie. They routinely compared American women scientists of all ages to Curie and, finding them wanting, justified not hiring them on the unreasonable grounds that they were not as good as she, twice a Nobel Laureate! (127)

What seemed to begin as an effort to raise women's aspirations in science resulted, not in a widening of opportunities for women, but rather in raising expectations "to almost unattainable heights." (127)

Rossiter recognizes the similarity of this phenomenon to the treatment of Somerville but does not fully explore the significance of the similarity. In 1875, Maria Mitchell had been told by a rather liberal college president that he "would hire a woman scientist if she was as good as Mary Somerville." (64) Many of the comments quoted above support the contention that the strategy of using exceptional women to create increasingly higher standards was being used much earlier than the 1920s, when it first began to be used in relation to Curie. I believe it could perhaps be more appropriately called "the Mary Somerville strategy," because Somerville's case offers one of the earliest and most significant examples of its use. The agitation for establishing women's opportunities in higher education that began during Somerville's lifetime reached its most intense moments in the years following her death. In some ways, Somerville's example was more powerful than Curie's in that she not only set an extremely high standard, one that could only be achieved with an unusual combination of ability, perseverance, and luck, but she also could be used to reinforce non-competitive norms of womanhood in a way that Curie could not.

Somerville's story indicates here, as in other places, how much variation there can be within a general pattern of excluding women from science. It also demonstrates how interesting it can be to live long enough for significant historical changes to occur. The task of the rhetoric of science during the height of Mary Somerville's scientific career was to associate science with virtue and stability, and Somerville served that purpose well. But the rhetoric of science does not stand still for long and soon moved on to projects for which the example of Mary Somerville could not provide support. The educational goals that emerged from the interpretations of Somerville's life suggested that women should be educated to make them better at the same role, not to give them another role. As the rhetoric of the revolutionaries became more adversarial, the response of the opposition became even more forceful.[8] Somerville was a venerated figure, but she was only one woman, and she could not easily serve as an example for an activist women's movement that sought explicitly to emancipate women from the domestic sphere.

The Nature of Her Story

One subtle but significant aspect of Somerville's forgetting arises from the nature of her life story, a tale in which there is nothing lurid, sensational, tragic, or shocking, even in the unexpurgated version. The idealized account presented in *Personal Recollections* was remarkably close to the reality of Somerville's life, but it lacked the intrigue that derives from notoriety. Mary Somerville was notable but not notorious. When Dorothy Sayers, herself an alumna of Somerville College, created a fictional counterpart of the school, which she called Shrewsbury, the character for whom the college was named was of an entirely different kind than Mary Somerville, virtually an antithesis. Mary, Countess of Shrewsbury, in whose honor the college had been named, is described as an

> ominous patroness . . . a great intellectual, indeed, but something of a holy terror; uncontrollable by her menfolk, undaunted by the Tower, contemptuously silent before the Privy Council . . . a lady with a turn for invective remarkable even in an age when few mouths suffered from mealiness. She seemed, in fact, to be the epitome of every alarming quality which a learned woman is popularly credited with developing. Her husband, the "great and glorious Earl of Shrewsbury," had purchased domestic peace at a price; for, said Bacon, there was "a greater than he, which is my Lady of Shrewsbury." (1968: 47)

Somerville's story also differs radically from that of two other scientific women with whom she might be compared, Emilie du Châtelet and Ada Byron Lovelace. Du Châtelet produced the first complete translation of Newton's *Principia* from English into French. This work was not published until after her death, and most of her notoriety during her lifetime arose from her association with Voltaire and from a variety of unconventional behaviors. After her affair with Voltaire ended, Emilie found herself pregnant by a younger man. She was forty-

---

[8] This is asserted in Paulina Davis' letter to Somerville, when Davis asked Somerville to write a letter to an American suffrage conference. (Somerville Collection, Folder MSDIP-15 in Dep.c.376.)

three years old, and in the midst of completing her translation of the *Principia*. She worked feverishly on the project during her pregnancy, writing to the father of her child, "I must do this . . . or lose the fruit of my labours if I should die in child-bed . . . I finish it from reason and honour." (Du Châtelet quoted in Alic 1986: 147) Du Châtelet died – tragically but not atypically for a woman of the eighteenth century – a few days after the birth of her child.

Ada Byron Lovelace is perhaps best known as the daughter of one brilliant but ultimately tragic man, Lord Byron, and as the protégé of another, Charles Babbage. Lovelace was also a friend of Mary Somerville. (PR 154). The two corresponded on a variety of mathematical and scientific subjects and often went together to Babbage's home to view his analytical engine or to meet with the array of scientific luminaries who gathered there. Ada plays a large role in Maboth Moseley's biography of Babbage, *Irascible Genius* (1964). Despite the early promise Ada Byron showed, she met with a tragic end. She died a very painful death from cancer at the age of 36, the same age at which her father had died. She had spent much of her adult life embroiled in financial and legal difficulties arising from her attempt to use a mathematical system to bet on horse racing. She was passionate, indeed immoderate, in her interests and wrote to Mary Somerville describing her "great anxiety" about Babbage's calculating Machine. "I am afraid," she wrote on July 8, 1834, "that when a machine, or a lecture, or anything of the kind, comes in my way, I have no regard for time, space, and any ordinary obstacles." (Somerville Collection, Folder MSBY-2 in Dep.c.367)

Ada Lovelace's correspondence with Woronozow Greig suggests that her marriage to Lovelace, who had been a college friend of Greig's, was never a happy one, and that she was equally unhappy being a mother. In letters written on February 4 and 12, 1845, she wrote to Greig, "Unfortunately, *every* year adds to my utter *want* of pleasure in my children. They are to me irksome duties and nothing more! . . . I am satisfied that *no* man *could* satisfy *my* ideal." (Somerville Collection, Folder MSBY-9 in Dep.c.367) Moseley sees Ada Lovelace as a genius, and seems to view the unconventional and unfortunate aspects of Lovelace's life as being consistent with the genius she possessed. Reading the accounts of Du Châtelet, Lovelace, and the fictional Lady Shrewsbury, one begins to get the sense that Mary Somerville was somehow too normal, too sane, too happy, and too well adjusted to be a genius. Geniuses are not usually presumed to live to their ninety-second year in the full retention of their intellectual powers and die quietly in their sleep with their devoted children around them, at peace with God and all the world. They are not imagined to make the kinds of statements Somerville made in the closing pages of *Personal Recollections*.

> The Blue Peter has been long flying at my foremast, and now that I am in my ninety-second year I must soon expect the signal for sailing. It is a solemn voyage, but it does not disturb my tranquillity . . . profoundly grateful for the innumerable blessings I have received. . . . I have every reason to be thankful that my intellect is still unimpaired . . . my daughters . . . make the infirmities of age so light to me that I am perfectly happy. (PR 373–374)

Such a life is perhaps ideal for living but not nearly as interesting to read about; it might be described as more enviable than engaging. In the end, it seems that the very conformity and normalcy on which her fame to a great extent depended ultimately made Somerville a less likely subject for the kind of women's history that implicitly discourages women from public intellectual pursuits by associating them with unfortunate consequences, usually incurred by a failure to adhere to ideals of feminine domesticity.[9]

Outside the Patterns of History

Mary Somerville's treatment as a woman in science relates to one of the most important generalizations about her being forgotten – she fell outside the patterns of history. Her status as a woman played a critical role in that process, but there were other factors involved as well. In his *History of the Inductive Sciences* (1837), Whewell established the pattern of organizing history of science around major "original" discoveries. It is a testimonial to the esteem in which Somerville was held by her contemporaries that Whewell was taken to task for not including Somerville in his *History*. The critical reviewer was David Brewster writing for the *Edinburgh Review*. (Brewster 1837) In a letter to the Reverend Richard Jones written September 6, 1837, Whewell explains the difficulty in his position.

> For what am I to say? I have not mentioned Mrs. Somerville. There was no pretext for mentioning her in the way of original discovery; and I am not to compliment away the character of impartiality which alone can give value to a history. But it is harsh and pompous to say this, and I had rather leave people to think it. (Whewell in Todhunter 1876, 2: 260)

---

[9] It is interesting to recall that Hypatia and Maria Agnesi, the two other women Whewell honored along with Somerville, also met with unfortunate ends and that neither conformed entirely to conventional domestic ideals.

Whewell clearly recognized the value of Somerville's contributions and her status within the community, but he was wedded to and helped establish a notion of the history of science that was organized around dramatic discoveries. As his review of *Connexion* demonstrated, he went to great lengths to make a conceptual space for Somerville in the context of contemporary thinking about gender and science. But he was not willing to create a space for her in history. In the subsequent years, the history of science began to move in directions that would eventually seem to obscure the possibility that such a space could ever have existed.

Somerville was included in William Chambers's *Stories of Remarkable Persons* (1878) and W. H. Davenport Adams's *Celebrated Englishwomen of the Victorian Era* (1884) but not in J. T. Merz's *A History of European Thought in the Nineteenth Century* (1904–1912), Sedgwick and Tyler's *A Short History of Science* (1919), Harvey-Gibson's *Two Thousand Years of Science* (1929), Dampier's *A History of Science and Its Relations with Philosophy and Religion* (1930), Singer's *Short History of Scientific Ideas to 1900* (1959), Mason's *A History of the Sciences* (1962), *The Cambridge Illustrated History of the World's Science* (1983), or in the list of "Scientists, Inventors, and Engineers" included in *100 Great Nineteenth-Century Lives* (1983).[10] She has appeared routinely in treatments of women in science and occasionally in the journals of scientific professional societies. She seems to have been given slightly more attention during the era when women first began to receive degrees at Oxford but was presented then primarily as a model of what self-help can accomplish and as one whose life yielded no suggestions for social change. She ultimately seems to have been interpreted more as an important figure in the history of Victorian culture or the history of women than as an important figure in the history of science.[11]

In the final analysis, it seems clear that gender played a crucial role in the way Somerville was to be remembered, but there were other factors. The most important of these is the fact that she was credited with no great discovery. There was no theorem, unit of measurement, principle, or law associated with her name, no accomplishment that all

[10] The preface to Canning's book reflects the uncritical attitude typical of much historical work with relation to women as he explains why there are only ten women among the one hundred people chosen: "Mothers would seem to be much more important than fathers in the training of the young; many great men lived their public lives in permanent awe – like Napoleon – of Madame Mère. Yet in the last century anyway it is men rather than women who emerge to greatness, or at least that form of it that merits public notice and acclaim." (Esmond Wright, "Introduction," 9–10)

[11] This inclusion in the histories outside of science may have helped diffuse any public sense that Somerville was not given due credit and respect.

students of science, mathematics, and engineering would learn in the course of their studies. This made it much easier for historians to ignore her and much less likely that scientists, mathematicians, or engineers would know who she was. There was little chance that historians or practitioners would stumble across her name and wonder who she was and what exactly she contributed.

Another significant factor in her being forgotten was that she came to prominence in a period when the reputation-building power of science was concentrated within a relatively small group of people who took a unitary approach to intellectual pursuits and were not yet discipline bound. As we have seen, this group was in many ways much easier for a woman to become a part of than the university faculties and professional organizations of the later nineteenth century would be. Somerville's reputation was founded on first-hand acquaintance with those who held reputation-building power. She had been acknowledged by Laplace and assimilated into the scientific elite long before she had published anything of lasting significance. These personal contacts were invaluable in building her career and reputation, but her reputation began to die when the people into whose circle she had been initiated began to die.

The Legacy of Somerville College

Were it not for the existence of Somerville College, Oxford, the name of Mary Somerville would be familiar only to a handful of historians. The college is the only high-profile memorial to Mary Somerville and is an important part of the story of Mary Somerville as a public and historical figure. The two major historical treatments of the college are *Somerville College: 1879–1921* by Muriel St. Clare Byrne and Catherine Hope Mansfield (1921) and *Somerville for Women: An Oxford College 1879–1993* by Pauline Adams (1996). The figure of Mary Somerville plays a much larger role in the latter than in the former. In both histories, Mary Somerville is portrayed not so much as a guiding presence as a compatible but loosely construed role model, perhaps because Mary Somerville was just the first of a long line of distinguished women associated with the College.[12] Overall, the identity the college assumed is consistent with Somerville's own ideals and outlook, but it seems to have evolved independent of any detailed sense of who Mary Somerville really was.

[12] These women included Margaret Thatcher and Indira Ghandi.

Byrne and Mansfield rarely mention Mary Somerville, and, when they do, her role is somewhat minimized. The following comment both reflects the multifaceted identity Somerville herself embodied and illustrates the tendency to minimize her as a role model: "One fact alone clearly emerges, and that is that there is no typical Somervillian who can be put forward as a miniature Mary Somerville. Somerville produces no definite type." (73) This failure to produce a definite type is seen as a positive feature that has

> ventilated the atmosphere of the College, and given its members a quick toleration of all points of view. Somerville herself is broad-minded, but it would be unfair to say that she has inflicted that rather negative quality on all Somervillians as their creed. The college turns out the orthodox and the unconventional, the athlete and the blue-stocking, the poetess and the social butterfly. (73)

In this passage as in others, Byrne and Mansfield tend to elide the personality of the college with the personality of the woman for whom it was named, or, rather, the college is often treated as a kind of person whose presence and attitudes are more palpable than those of the individual Mary Somerville. The authors' description of the college's philosophy and strategy for thriving in an all-male environment is also strikingly similar to those of Mary Somerville:

> Somerville College was born, then, of enlightened ideas, and in particular of a progressive conception of woman, which clothed itself discreetly in the hereditary feminine garb of modest manners and watchful tact. To live at the gates of a venerable and conservative University, with support and encouragement from some of its most liberal members, was to be perpetually conscious both of deference due to ideas of the past and the beckoning promises of the future. (15)

There is no indication of how aware Byrne and Mansfield were of the details of Mary Somerville's own life, but the alteration of a few words would make this passage a quite accurate description of Somerville herself, especially in its image of "a progressive conception of woman . . . clothed . . . discreetly in the hereditary feminine garb of modest manners and watchful tact."

Adams's history, which is longer and gives more extensive treatment to Mary Somerville as an individual, also delineates a number of traits shared by Somerville and the college that bears her name. Probably the most important of these is the college's nondenominational character. Adams is careful to distinguish nondenominational from "irreligious," and indicates that the founders of the college chose holding "all denominations in equal respect" over being "purely secular." (1996: 353) Their aim was not to devalue religion but to distance religion from education and to promote religious tolerance. This stance is entirely compatible with Somerville's own, especially as it is revealed in her autobiography, and remains to this day an important aspect of the college's identity.

A second of Mary Somerville's traits that seems to have been influential in defining the identity of the college is her dedication to science and her ability to combine high achievements in science with her role as a wife, mother, and cultivated woman. The college has placed particular value on science research and teaching; it was "the first of the women's halls to appoint a tutor in the sciences" (1996: 357); and it enjoyed the support of scientists and their wives, who helped women gain access to science lectures and laboratories. As Adams puts it,

> Somerville has taken pride in the fact that it was named after a woman scientist, that it could count a Nobel prizewinner for chemistry among its alumnae-fellows, and that for twenty-two crucial years after the second World War it had a distinguished scientist as Principal. Both Dorothy Hodgkin and Janet Vaughan were notable in their day for encouraging women to undertake scientific research, and to pursue their scientific work – as they had themselves, and Mary Somerville had before them – in conjunction with family responsibilities. (357)

This dedication to science and to encouraging women to live what might be called a complete life – a life that combines the rewards of a career with the rewards of traditional womanhood – has clearly been an important part of the college's self-image and may well have been one of the keys to its success.

Both Somerville and the women who followed her found that living a complete life was not easy, though they varied in the degrees to which they were willing to acknowledge the difficulty. It seems that many Somervillians have also felt the kind of pressure to excel that Mary Somerville felt once she had established herself among the scientific elite. Adams explains, "The conspicuous achievement of some of the college's alumnae has tended to engender a particularly Somervillian neurosis: that unless one has raised a family and become a Fellow of the Royal Society by the age of fifty . . . one has somehow failed to live up to the college's expectations." (363) At the height of her success, which she achieved at the somewhat advanced age of nearly fifty-one, Somerville shared this same feeling of having failed to live up to expectations. It seems doubtful to me that these similarities between Somerville and alumnae of the college grow out of detailed knowledge of the specifics of Mary Somerville's life. It seems more likely that the similarities reflect recurring themes in the lives of intellectual women.

As previous chapters have demonstrated, Mary Somerville felt strongly about the cause of women's education. The record suggests that Somerville College and the Somervillians succeeded in achieving some of the goals that were most important to Mary Somerville. They demonstrated that higher education does not "ruin" women for family life, that well-trained women can hold down careers, and that "celibacy is no longer the necessary price of a successful career." (363) This aspect of the Somerville tradition seems to have provided an inspiration for women students and may also have served as a self-fulfilling prophecy of the kind that has often been important in the history of intellectual women. My knowledge of Mary Somerville leads me to believe that she would have been very pleased to have had her name associated with and to have provided encouragement to such an enterprise. Through institutional astuteness and its encouragment of young women to follow their own genius, develop their abilities to the fullest, and define their own identities, Somerville College provided for many women the social and psychological space for which Mary Somerville struggled and that the majority of her contemporaries never achieved. Given how "kept down" she often felt, especially as a very young woman, she would probably have found it more significant to have helped advance the institutional opportunities for self-determination of other intellectual women than to have been strictly construed as a role model.

One of Adams's most interesting arguments about the history of the college is that "Somerville has always had a powerful sense of its own identity, as is perhaps to be expected in an institution which was founded by a splinter-group in defense of a principle. But the college has retained . . . [a] hold on the public imagination . . . [a] reputation for producing formidable women." (1996: 1) The stories of Somerville the college and Somerville the woman both demonstrate the power of intellectual women to capture the public imagination. They also demonstrate that history does not stand still. In 1992 after extended and sometimes contentious debate, Somerville College ceased to be a women's college and began to admit men. It remains to be seen whether the legacy of Mary Somerville will be as meaningful, inspirational, or memorable to the current and coming generations of Somervillians as it was to previous ones.

Reconsidering the legacy of Mary Somerville makes it clear how limited many common perceptions about the history of science and the role of women in it are and offers a much richer and more finely articulated view of what science involves. Although these limitations and the potentially richer view that comes from getting beyond them are well recognized by specialized historians, they are much less widely recognized beyond that group. And, in any case, the particularities of Somerville's life and work add depth and dimension to the new historical view that is emerging.

Women at the Apex of the Prestige System in Science

One of the central characteristics of traditional history of science is that it has operated with a de facto canon, that is, a well-recognized group of key figures who exemplify important values and often take on a legendary quality. All candidates for historical significance must meet the criteria that have traditionally been used to establish significance. In a sense, to be famous in science is to *have been* famous, so that the history of science carries the burden of its own history.

The effects of the historical circumstances that inhibited the participation and recognition of women in science cannot be undone altogether, but it is possible to look at history in a way that makes the contributions of women easier to see and appreciate. Mary Somerville demonstrates that even in the nineteenth century, it was possible for a woman to attain eminence in science. As such, she offers one of our best hopes for remapping the territory entirely, so that it would seem incomplete without her and other women scientists and without a consideration of the kinds of things that she and other women did.

In her treatment of "women at the apex of the prestige system of science," (xii)[13] Hilary Rose (1994) characterizes Somerville as a "brilliant scientist" treated badly (115), and there is certainly justification for this view. But it is also important to illuminate the extent to which Somerville was able to push the boundaries of what was considered possible at the time. Rose's analysis focuses on the Royal Society and the Nobel Prize and thus deals with a system that was configured differently than the one Somerville operated in during the nineteenth century. Given appropriate consideration of these differences, I believe the record justifies the conclusion that Mary Somerville was at the apex of the prestige system of science in her day.[14]

---

[13] For discussion of women and the Royal Society, in addition to Rose (1994), see Mason (1992, 1995, 1996).

[14] Nobel Prizes were not awarded until the beginning of the twentieth century, and the Royal Society did not elect its first women fellows until 1945, even though a legal opinion rendered in 1922 made it clear it was no longer legally tenable to exclude women. (Rose 1994: 137, 116) My conclusion also presumes that, as with Nobel Prizes, there is more than one position to be occupied at the apex, i.e., that the apex is not a single position.

In Somerville's era, power in science was not yet concentrated in the way it would be by the time of Nobel Prizes existed. Moreover, her era was characterized not by a single dominant figure such as Newton or Darwin, but rather by a network of activity and influence, institutions and individuals. The net result was that the power associated with science was relatively diffuse. She had close ties to all of the influential cultivators of science, most notably John Herschel, Charles Babbage, Michael Faraday, Laplace, and William Whewell, and was well integrated into the networks of power connected with her work. It would have been inconceivable for a woman to have become *the* central figure in science in her day, but also there *was* no central figure.

Like the Nobel laureates in Rose's analysis, Somerville functioned as an icon "of the fusion of scientific knowledge and cultural power." (Rose 1994: 138) The obituaries and reviews cited earlier, the poetic responses Somerville evoked, and the way she was treated by advocates of women's rights provide evidence of the cultural capital Somerville possessed. Like the first women elected to the Royal Society, she received forms of recognition previously unavailable to women and had her work "weighed on equal terms" with that of men (at least in the sense that it was judged by the same standards as those used for men, though with closer scrutiny because she was female). Both published and private commentary on Somerville by other scientists is characterized by a pervasive sense that she was treated as a respected equal and integral member of the community of cultivators of science. This is clear when these other scientists praise her work and perhaps even clearer when they criticize or take issue with her.

Mary Somerville enjoyed the maximum equality that could possibly have been expected at the time and occupied a place that was somehow more visible because she was the only woman to occupy such a space. Given the times in which she lived, it would have been unthinkable for her to have been elected as a fellow of the Royal Society. On the other hand, the men who recognized Somerville did not seem to feel that it was inevitable that they would have to recognize her, and their recognition seems to be based on genuine admiration for her and her work.

It is no less true, however, that these same men were part of institutions that showed considerable skill and success overall at resisting and dealing with women. They handled her rise to eminence in a way that did not undermine the male-dominated character of science. To use Rose's term, her admission was "man" aged. They found ways of honoring her other than formal admission to membership – medals, honorary membership, a bust commissioned – alternatives which were quite extraordinary in their own way. The net effect was that the recognition Somerville received was handled so that it promoted the scientific enterprise without challenging its male identity. And though Somerville's example eventually seems to have had a positive impact on the educational opportunities of women, the full realization of that opportunity was a long time in coming.

One of the most striking things that emerges from Somerville's story is the portrait it offers of the power, tenacity, and adaptability of gendered thinking, a set of ideas configured so that it could keep women in subordinate positions and diffuse the tensions and contradictions that inevitably arose. Still, in the end, it seems to me that Somerville meets the traditional canonical criterion of *having been* famous and that a consideration of her career highlights the opportunities for enriching the values and the view that have been used in identifying historical significance.

The Character of Scientific Discourse

Outside the group of scholars concerned with the history of scientific communication and the area known as "science as public culture," the most common view of scientific discourse is rather narrow and has tended to assume that true scientific discourse is highly specialized and epitomized in the case of scientists writing up their own experimental or theoretical work. Within this view, all other forms of discourse tend to be seen as peripheral to science and to be categorized as either "popularization" or "exposition." There has been an almost universal tendency to create a hierarchy in which popularization is seen as a distorted or inferior form of specialist discourse. (Myers 1990: 141; Shinn and Whitley 1985)

While it is undoubtedly true that some forms of discourse related to science have been more important than others for developing new techniques and theories and for validating new scientific knowledge, it is no less true that the authentication and interpretation of knowledge are *processes* in which many different kinds of discourse are involved. One of the great strengths of Somerville's career is the way that it expands and enriches our view of the character of these processes and the multiplicity of scientific discourses. It shows how profoundly limited most conceptions of popularization are. For example, her work clearly demonstrates the inadequacy of a notion of "primary" and "secondary" forms of discourse. While it might be viewed as a good illustration of the notion of an expository continuum (Cloître and Shinn 1985), I believe her work is best understood as an illustration of the existence of a network of overlapping, interrelated, and often intertwined discourses – what might be termed a network of scientific discourse, or discourse network. The breadth of subjects she dealt with and the range of audiences she addressed meant that her presence within the discourse network was quite extensive. Her work is particularly useful for highlighting a frequently overlooked aspect of the network: the audiences constituted of experts in fields and disciplines other than the one in which new knowledge originated.

One of the most striking things her writing demonstrates is that authentication and interpretation *are* processes, in the sense that they involve stages and steps. Especially if her works are considered in the order she wrote them, it is easy to get a clear sense of the process by which new scientific knowledge radiates outward from its point of origin and becomes incorporated into increasingly larger systems of knowledge and belief, so that it begins as part of a working model of a particular phenomenon and eventually is incorporated into a shared view of the world. Along the way, its connections with other phenomena, along with its practical and philosophical implications, are delineated.

The significance of new observations or experimental results is often not apparent until a very large number of these have been accumulated and a pattern emerges. Much of Somerville's work was concerned with charting the accumulation as it occurred, articulating new patterns and connections as they emerged, and assessing where various theories stood in the process of validation and acceptance. In her role as intellectual, she also helped establish the larger meaning and significance of highly specialized research. She demonstrated the ways in which new findings could be accommodated into existing belief systems and also indicated ways in which a particular world view would have to be renovated to provide room for the conclusions supported by new research. Although she embraced the myth of original genius on occasion, and judged herself harshly in light of it, she also viewed progress in science as a collective, social process in which certain individuals became the articulators of new points of view that were already at work in the minds of many. Although she did not establish herself as one of the great synthesizers in the mold of Newton, she played a significant role in articulating and mapping "the collective mind," especially as it related to science.

One way to characterize all of these activities is to look at them as multiple acts of interpretation. The importance of interpretation in science is perhaps best illustrated in Somerville's handling of the atheistic implications of Laplace's ideas and in the role she played in what might be termed the process of converting the heretical into the commonplace. As the analysis in chapter 3 demonstrated, the interpretive framework Somerville created in the "Preliminary Dissertation" located Laplace within a theistic framework. That framework did not contradict or interfere with the technical quality (i.e., the quantitative or analytical aspects) of Laplace's work, but it did very strongly associate that work with an established aesthetic tradition that demonstrated its compatibility with, indeed its support for, a theistic world view. In this way, the rhetoric of science she created played a significant role in facilitating one of the major transitions that occurred between the late eighteenth and late nineteenth centuries. Henry Smith Williams outlined this transition in a book entitled *The Story of Nineteenth Century Science*, which he published in 1901. Looking at the changes that occurred, he says:

> Mark the transition. In the year 1600, Bruno was burned at the stake for teaching that our earth is not the center of the universe. In 1700, Newton was pronounced "impious and heretical" by a large school of philosophers for declaring that the force which holds the planets in their orbit is universal gravitation. In 1800 Laplace and Herschel are honored for teaching that gravitation built up the system which it still controls; that our universe is but a minor nebula, our sun but a minor star, our earth a mere atom of matter, our race only one of myriad races peopling an infinity of worlds. Doctrines which but the span of two human lives before would have brought their enunciators to the stake were now pronounced not impious but sublime. (16–17)

Williams has sacrificed some precision in his dates for rhetorical effect; nevertheless, his analysis captures the radical changes in world view and the way new scientific findings in particular are valued and interpreted. The rhetoric of science that Somerville borrowed from the scientific poets and adapted for scientific prose played a significant role in the transition through which scientific doctrines were transformed from the impious to the sublime. A major aspect of this transition was a large number of acts of interpretation and reinterpretation.

It is important to recognize that the business of translating scientific knowledge into broader world views was not the exclusive province of religious people. George Basalla has compared Somerville to T. H. Huxley, indicating that she was less notorious but equally influential. (Basalla 1963: 531) At first this comparison might seem inappropriate given the very different philosophical orientations of Huxley and Somerville, but a deeper examination reveals some important similarities. Both were actively involved in disseminating and interpreting new science and elucidating its significance for human belief systems. Careful analysis of belief and unbelief in Victorian Britain has revealed the ways in which the rhetorical and intellectual transitions to which

Somerville contributed paved the way for the acceptance of Darwinism and a secular world view. (Lightman 1987)

Somerville's work also provides an opportunity to examine the power of weaving together forms of discourse that are presumed to be either separate or in tension. Analysis of her work highlights the ways in which science and aesthetic pleasure are intertwined and mutually supporting rather than adversarial or simply complementary. Alexander von Humboldt, the Prussian explorer, naturalist, and scientist, expressed such a hope that work conveying empirical or scientific knowledge might have lasting aesthetic and literary value.

> It has frequently been regarded as a subject of discouraging consideration, that while purely literary products of intellectual activity are rooted in the depths of feeling, and interwoven with the creative force of imagination, all works treating empirical knowledge, and of the connection of natural phenomena and physical laws, are subject to the most marked modifications of form in the lapse of short periods of time [and] continually being consigned to oblivion as unreadable. . . . I would . . . venture to hope that an attempt to delineate nature in all its vivid animation and exalted grandeur, and to trace the *stable* amid the vacillating, ever-recurring alternation of physical metamorphoses, will not be wholly disregarded even at a future age. (Humboldt 1850: xi–xii)

Humboldt was, of course, describing his own project in *Kosmos*, a work that was often compared with *Physical Geography*, but he might have been describing Somerville's work as well.

Humboldt's life and work form an interesting comparison with Somerville's because of the similarities in the projects they undertook and because Humboldt devoted a great deal of thought and energy to articulating the aims, motivation, and method of that common enterprise.[15] He is particularly interesting as a theorist on the relationship of aesthetics to science, an area in which Somerville was deeply involved and exceptionally skillful. He and Somerville shared a multifaceted view of science as being at once aesthetic, utilitarian, and morally progressive. Despite these similarities in sentiment, there were some important differences. Humboldt did not share Somerville's religious beliefs though he, too, rejected Darwinism. Perhaps most significantly,

---

[15] Though it is probable that they met only once, Humboldt and Somerville shared a number of similarities in life, method, and style. Humboldt was a scholar of broad interests and acquirements who had an interest in practical affairs and politics as well as science. Both he and Somerville lived long lives, active to the end, producing successive editions of works in order to keep them up to date; both embraced a belief in science as a collective activity; and both were recognized as self-effacing, free from jealousy, and generous in their efforts to help others.

the independence afforded by his wealth, his sex, and his freedom from family obligations enabled him to undertake research and travel on a scale that would have been impossible for Somerville.

Confined as she was by gender and economic circumstances, she could never have fully earned the title of "a female Humboldt," but she did earn Humboldt's admiration, recognition, and respect. In a letter of July 12, 1849, thanking Somerville for a copy of *Physical Geography*, Humboldt praised Somerville's superior combination of precision, lucidity, and taste; her high attainments in mathematical analysis; and the comprehensiveness of her knowledge. He found *Physical Geography* both charming and instructive. "You alone could provide your literature with an original cosmological work, a work written with the lucidity and taste that distinguish everything that comes from your pen," he told her, "I know of no work in Physical Geography in any language that one could compare to yours." (PR 287–288)[16] When he was visited by the American astronomer Maria Mitchell in 1858, Humboldt spoke of *Physical Geography* "with admiration . . . said it was excellent because so concise." (Quoted by Mitchell in Kendall 1896: 166) "A German woman," Humboldt continued, "would have used more words."

Neither Humboldt's nor Somerville's scientific writing has been able to overcome in any significant way the stigma of "antiquity" to which Humboldt alluded. Still, both refute Edmund Burke's assertion that "it is our ignorance of natural things that causes all our admiration, and chiefly excites our passions." (Quoted by Humboldt 1850, 1:40) The situation, contended Humboldt, was exactly the opposite. Both aesthetic pleasure and the progress of science grew out of the perception of unity, of seeing the universe as an interconnected whole. Somerville's gift for conciseness made it in some ways easier for her to convey that wholeness and to show the ways in which the aesthetic and intellectual pleasures of science were linked. Where Humboldt tended to explore the aesthetic dimensions of science in works that were separate from his quantitative, analytical works, Somerville blended the two discourses into a seamless whole, with scientific content dominating and the aesthetic element adding power, meaning, and pleasure.

If Somerville's work cannot be classified as "purely literary," it also does not conform to the norms of the "scientific" discourse as these are strictly construed. The value placed on Somerville's writing by the scientific experts of her day supports the need for a multiplicity of scien-

---

[16] I am grateful to Melvin Cherno for his assistance with the translation of these passages from Humboldt's letter to Somerville.

tific discourses. James Clerk Maxwell took up these issues in an address to the mathematical and physical sections of the BAAS on September 15, 1870, just a few years before Mary Somerville's death. Maxwell begins by describing the mental disposition underlying what many people understand as the purest form of scientific discourse.

> There are some men who, when any relation or law, however complex, is put before them in symbolic form, can grasp its full meaning as a relation among abstract quantities. . . . The mental image of the concrete reality seems rather to disturb than to assist their contemplations. (1890: 219)

Maxwell concludes that "the great majority of mankind are utterly unable, without long training, to retain in their minds the unembodied symbols of the pure mathematician" and that science will never "become popular, and yet remain scientific." (219)

He then goes on to describe other modes of contemplation and discourse.

> There are others who feel more enjoyment in following geometrical forms, which they draw on paper, or build up in the empty space before them.
>
> Others, again, are not content unless they can project their whole physical energies into the scene which they conjure up. They learn at what rate the planets rush throughout space, and they experience a delightful feeling of exhilaration. They calculate the forces with which the heavenly bodies pull at one another, and they feel their own muscles straining with the effort.
>
> To such men momentum, energy, mass are not mere abstract expressions of the results of scientific inquiry. They are words of power, which stir their souls like the memories of childhood. (220)

Maxwell then draws out the implications of this analysis:

> For the sake of persons of these different types, scientific truth should be presented in different forms, and should be regarded as equally scientific, whether it appears in the robust form and the vivid colouring of a physical illustration, or in the tenuity and paleness of a symbolical expression. (220)

Maxwell is not positing a single "scientific" mind or defining it as the one which grasps "the tenuity and paleness" of the "symbolic expression" ordinarily presumed to be the epitome of scientific communication. Rather, he envisions a multiplicity of discourses, each designed to bring science within the grasp of a differently configured intellect. His description of these discourses reflects the role that imagination, drama, visualization, and aesthetics would play in them. As the analysis of Somerville's writing presented in this book has shown, Somerville excelled in producing "vivid," "robust," and "soul stirring" accounts that also conveyed the detailed substance of science.

It is tempting to dismiss the kind of writing Somerville did as the ideal of an earlier era that can no longer be achieved because of the rate at which knowledge has expanded and continues to expand. While it is probably true that it is no longer possible to command the whole range of knowledge Somerville possessed, one of the most important objectives of her work is still attainable and seems increasingly desirable. That objective, as articulated by Stefan Collini in his recent introduction to C. P. Snow's *Two Cultures*, is "to impart to the non-specialist some sense of the significance if not the detail of extremely technical research." (lvix) As Collini notes, this is a matter of skill and desire to understand how specialized "activities fit into a larger cultural whole." (lviii) Such activities, which require both technical expertise and rhetorical skill, carry relatively little value or prestige today, but they were crucially important for the cultivators of science of Somerville's day as they sought to facilitate meaningful cultural dialogue about the character and value of science.

Gender as It Relates to Models of Thought and Perception

One of the major themes of this book has been the association of science with illumination and enlightenment. Somerville functioned as both embodiment and purveyor of a morally informed, aesthetically rich scientific illumination. This illumination entailed both the literal ability to see what others do not see and the figurative state of enlightenment, of having one's conceptions enlarged, of being informed in an uplifting way, and of being freed to operate intellectually and morally on a higher level. This illumination united two modes of thought and perception into something much more powerful than either alone. To use a typically nineteenth-century characterization, it combined "quickness of perception, strength of attachment, and brilliance of imagination" with the breadth and power of "abstract knowledge" acquired through "strong and vigorous exertion." (Hale 1840: 273)[17] It thus brought

---

[17] These characterizations come from an article entitled "Comparative Intellectual Character of the Sexes," which was written by Mary Hale for *Godey's Lady's Book*. The article describes Somerville as part of "a living galaxy of merit. . . . unremitting study and investigation" (274), but says little about her otherwise. It provides a window into

together the strengths of both modes and overcame the limitations associated with typically masculine and typically feminine modes. This hybrid mode might be conceived as transcendent feminine. Throughout, this book has attempted to provide a rich view of the range and character of the responses Somerville evoked and to account for those responses. Although Somerville did not theorize about her own craft as a presenter of science, she did in *Physical Geography* offer a general model for the evocation of aesthetic responses that is helpful in understanding how she worked. As she developed this theory, she articulated her own version of the theme *ut pictura poesis*, which explores an analogy between painting and poetry:

> Painting, like poetry, must come spontaneously, because a feeling for it depends upon innate sympathies in the human breast. Nothing external could affect us, unless there were corresponding ideas within; poetically constituted minds of the highest organization are most deeply impressed with whatever is excellent. All are not gifted with a strong perception of the beautiful, in the same way as some persons cannot see certain colours, or hear certain sounds. Those elevated sentiments which constitute genius are given to few; yet something akin, though inferior in degree, exists in most men. Consequently, though culture may not inspire genius, it cherishes and calls forth the natural perception of what is good and beautiful, and by that means improves the tone of the national mind, and forms a counterpoise to the all-absorbing useful and commercial. (PG 468)

Extending the analogy to include the scientifically enhanced perception of the natural world provides a theoretical perspective from which Somerville's work can be profitably viewed.

Rephrased, the analogy would be something like this: In science as in the arts, there are a few individuals equipped with strong perception and possessing the elevated conceptions that constitute genius. These individuals expand the perceptions of the majority and thus facilitate general enlightenment. They accomplish this by exploiting innate sympathies and seeking out the corresponding, commonly accepted ideas that will allow the public to be affected, or moved, by new principles. This view seems to have been, at least implicitly, the one with which Somerville worked in her own writing, as she portrayed science as an encounter with the good and the beautiful as well as the useful.[18] She allowed her readers both the privilege of a larger

early nineteenth-century debates regarding gender and intellect and an interesting point of comparison with Somerville's own ideas about women and intellect.

[18] It is interesting to note how remarkably similar this image is to Whewell's image of the passing of the torch of knowledge.

view and the opportunity to connect with something ambiguously but powerfully conceived as "higher."

Somerville's writing thus exploited the new possibilities offered by science for locating the reader in relation to her subject. This ability to conduct readers to sublime truths derived in part from her ability to connect the cosmic with the everyday and to present appealing images of the concrete reality revealed through science. Her accessibility grew out of an ability to present complex material in a way that put it within the reader's grasp without diminishing its illuminating potential. The following comments from Somerville's friend Mary Berry, written on September 18, 1834, in response to the "Preliminary Dissertation," reflect how this new relation affected a reader.

> Humbled, I must be, by finding my own intellect unequal to following beyond a first step, the explanations by which you seek to make easy to comprehension the marvellous phenomena of the universe – humbled, by feeling the intellectual difference between you and me. . . . yet informed, enlightened, and entertained with the series of sublime truths to which you conduct me. (PR 207)

A great deal of Somerville's success as a writer was based on this distinctive strategy for reaching non-specialist readers. She did not bring the material down to the level of the reader. Rather, she brought the reader up to the level of the material and provided the reader with a vision of the subject that made the effort seem worthwhile. She did this largely through an aesthetic tradition that portrayed the universe not as cold, colorless, and mechanical but as vivid, warm, and aesthetically satisfying – though scientifically understood nonetheless. Just as Herschel viewed the findings of Laplace as "a beautiful and animated comment on the cold and abstract pronouncement of the general law of gravitation," we might view Somerville's writings as "a beautiful and animated comment" on the gray and mechanistic picture of the physical world as it has often been assumed to be revealed through science.

The Usefulness of an Integrated Model of Women in the History of Science

Another major theme of this book has been the usefulness of an integrated model of the history of science that recognizes the many spaces within science that have been occupied and defined by women and that makes it easier to appreciate their creative responses to highly constrained opportunities.

Mary Somerville found herself in one such distinctive space in the scientific community of the early nineteenth century, a community that was still relatively undifferentiated intellectually within and whose boundaries were shifting and permeable. That space offered her the opportunity for the most advanced training in mathematical science that was available during her day. She took full advantage of that opportunity and used the exposure provided by a move to London and travels in Europe to expand her range of scientific contacts and expertise. First through face-to-face contact and then by reputation, she established herself as an accomplished scientist. She obviated confrontation, overcame resistance, and eventually succeeded in establishing herself as an equal by engaging the most distinguished scientists of the day as teachers and collaborators. When offered the challenge of presenting one of the most esoteric and recondite scientific works of the day to a broader audience, she responded creatively – not by producing two separate works for two different sets of readers but by producing a preliminary dissertation that framed the detailed aspects of the topic within a larger view of the history, aims, and methods of physical astronomy. Once she had established herself in elite science, she had greater room to operate. Responding to the very positive response the "Preliminary Dissertation" received and to her own intellectual engagement with the process of tracing connections among various discoveries in different branches of science, she created an even more innovative form of scientific writing – the comprehensive but concise synthetic survey of the state of knowledge in all of the physical sciences. She maintained and expanded her range of scientific contacts as she updated this survey through frequent editions over twenty-four years.

As the discipline of physical geography emerged, she took advantage of another opportunity and her own experience in creating synthetic accounts by producing the first treatment of the subject written in English. In the process, she shaped the definition and aims of a new discipline. As new developments in chemistry and biology opened up a whole new realm of natural phenomena on a microscopic scale, she put her considerable talents as a scientist and her powers as an imaginative writer to work in exploring the drama, complexity, and variety exhibited on the smallest of scales. These last works reflect the constraints and opportunities presented by a move away from Britain (and British science) and by age, which limited her physical mobility and dimmed her eyesight but left her intellect and imagination intact.

She took the opportunities offered, responded to a sense of what was wanted, made a virtue of the necessity of writing to earn money, and chose a spouse who both honored the space she made for herself and helped expand and link it to other spaces. She accepted all the claims that society imposed on a woman of her class, yet established herself as a scientist, an intellectual, and a public figure in a culture that conceived of women as private, irrational, and incapable of higher intellectual pursuits. She both accepted and resisted the boundaries imposed on women, and created in the process styles of writing about and conceptualizing science that were both innovative and highly valued.

One of the interesting aspects of her writing is her refusal to portray nature in any stereotypically gendered way. Her treatment of gender as it is read into the processes of nature gives play to the masculine and feminine but gives center stage to neutral, ungendered forces and entities. Most of the time, she talks about particular forces and phenomena that are not personified. Gravity and gravitation are "it." The heavens, the heavenly bodies, and the universe are "it" or "they." There is no evidence that she made conscious decisions about whether or how to gender nature, and she does not completely avoid it altogether. For example, both "God" and "man" are "He/he." The sun and the planet Jupiter are often "it" but sometimes "he." Venus is sometimes referred to as "her." The only strongly feminine entity in her scientific writing is the moon, which is almost always referred to as "she." This is noteworthy since the moon's relation to the sun has often been used as a metaphor for the relation of female intelligence to male intelligence. Perhaps most significant, however, is the fact that she does not personify nature or the earth as feminine, refer to the earth or nature as "she," or portray the relationship between humanity and nature as a male/female relationship. Nature with a capital "N" very rarely appears. On the few occasions when Somerville discusses the domination of nature, she speaks of "the subjection of some of the most powerful agents in nature to his [meaning man's] will." (PG 2) This suggests not a model of a weaker, feminine, passive nature dominated by a stronger, masculine, active science, but rather an engagement of worthy opponents. When she portrays the relationship between human beings and nature as a struggle (and she rarely does this), the story takes on epic proportions as an interaction of many actors and forces in which no single one can expect to gain a final victory or universal dominance.

This image, like many of the most powerful ones in Somerville's writing, embodies the principle she called counterpoise, the juxtaposition of forces moving in different directions. This principle is more complex and interesting than either balance or tension because it involves not just two opposing forces, but rather a multiplicity of forces acting in a variety of relations to each other. The principle of counter-

poise is what she saw in nature and attempted to capture in her portrayals of nature. It is also what I see in her. The notion of counterpoise
resonates strongly with emerging models for thinking about gendered
traits as they enter into our notions of human identity.[19]

Just as Somerville created a space for herself in the world of science,
I have attempted to create a larger and more well-defined space for her
in the history of science. Through the course of this book, the story of
Mary Somerville's life has been told several times: as a straightforward
factual account, as the story of how a timid but persistent young girl
became an eminent scientist, as the story of a talented and versatile
writer and intellectual, as the story of an object of public attention and
subject of historical memory, and as the story of an innovative woman
scientist who carved out a space of her own within the masculine world
of science. There is, I believe, no master narrative that would capture
the richness and complexity of Mary Somerville's life and work. My
original intent in bringing this work to a close was to recommit Mary
Somerville to memory, but I have come to believe that it would be more
fitting for her to become part of what Maxwell called "the living mind,"
as a reminder of the multifaceted, dynamic character of science and of
its power as a mode of perception and expression.

---

[19] In the world of ideas, dualisms seem increasingly inadequate, and reality seems more
readily characterized by principles like counterpoise. As Stefan Collini suggests, "We
need . . . something like multidimensional graph paper in which all the complex para-
meters which describe the interconnections and contrasts can be plotted simultaneously." (Collini 1993: v) Collini is speaking of academic disciplines, but I suspect that
this analogy holds not just for disciplines but also for individuals and the texts they
produce.

Epilogue

Science, Voice, and Vision

Gravitation . . . connects sun with sun throughout the wide extent of creation . . . every tremor it excites in any one planet is immediately transmitted to the farthest limits of the system . . . like sympathetic notes in music, or vibrations from the deep tones of an organ.

– Mary Somerville,
On the Connexion of the Physical Sciences

What Mary Somerville was and what she accomplished are captured in the self-portrait that serves as the frontispiece for this book. One of the most striking characteristics of the self-portrait is the directness with which the subject meets the gaze of the viewer. She deliberately presented herself as an author, pen in hand and seemingly in the midst of composition. Her clothing seems decidedly, perhaps even self-consciously, feminine. The face is portrayed with much more accuracy and detail than any other part of the portrait except for her hands. In fact, much of the background is treated in an impressionistic way that almost suggests that the portrait is not yet finished. For the style of portraiture of the time, the gaze is uncommonly direct. It is neither reluctant nor aggressive, but, rather, discerning, perspicacious, shrewd.

In one respect, the directness of the gaze reflects her voice as an author, the hard-won confidence and authority that Somerville enjoyed as a scientist and an intellectual. Her life story demonstrates what a great distance had to be traversed in order for her – or for any woman of her time – to overcome the norms of deference that she surmounted in meeting the viewer's gaze so directly. Writing about science helped her establish an authorial voice without moving too far outside conventional feminine behavior. Somerville did not obscure herself completely by creating a fictional world, but she also did not feature herself in her writing. The main characters in her accounts are the forces of nature, conceived as both multifarious and unified. She thus established a distinctive voice that preferred insight to ornament and embodied natural eloquence.

In another respect, the gaze represents one of the qualities that has been traced throughout this book – the power to illuminate, to bring things into view that have been hidden before. It embodies both Enlightenment and Romantic ideals. She looked at the world through her own eyes, trained by science and developed through experience, but also looked through the eyes of many others – both scientists and poets – to convey a collective vision of the world as it is – in all its drama, complexity, and vastness – and the world as it could be transformed through the transmission of enlightened ideas. She had mapped the world of science and effectively passed the torch of knowledge. In other words, she both defined what was known and conveyed to the world what it might mean in terms of a world view and the possibilities for human advancement.

This vision is clearly a holistic and multifaceted mode of perception, based on but not limited to the sense of sight. It is dynamic, played out through the descriptive patterns of the cosmic platform, tracing the mazes, nature as epic theater, and the living landscape. It is perhaps most succinctly expressed in the image of the universe as a musical instrument with gravity resonating through it "like sympathetic notes" or "vibrations from the deep tones of an organ." (Conn 1) It reflects a conception of the universe generated through a unique intersection of science, illumination, and the female mind.

Selected Bibliography

Abir-Am, Pnina G., and Dorinda Outram, eds. 1987. *Uneasy Careers and Intimate Lives: Women in Science 1789–1979*. New Brunswick, NJ: Rutgers University Press.

Adams, Pauline. 1996. *Somerville for Women: An Oxford College 1879–1993*. Oxford: Oxford University Press.

Adams, W. H. Davenport. 1884. *Celebrated Englishwomen of the Victorian Era*. London: White.

n.d. *Child-Life and Girlhood of Remarkable Women: A Series of Chapters from Female Biography*. London: Swan Sonnenschein.

Alic, Margaret. 1986. *Hypatia's Heritage: A History of Women in Science from Antiquity to the Nineteenth Century*. London: The Women's Press.

Ashfield, Andrew, and Peter de Bolla. 1996. Introduction to *The Sublime: A Reader in Eighteenth-Century Aesthetic Theory*. New York: Cambridge University Press.

Athenaeum. 1869. "On Molecular and Microscopic Science." Review of *On Molecular and Microscopic Science*. February 6: 202–203.

Athenaeum. 1873. Review of *Personal Recollections*. December 13: 761–762.

Atlantic Monthly. 1874. "Recent Literature." Review of *Personal Recollections*. (April): 489–502.

Basalla, George. 1963. "Mary Somerville: A Neglected Popularizer of Science." *New Scientist* 17: 531–533.

Bate, W. Jackson. 1970. *The Burden of the Past and the English Poet*. Quoted in Thomas Weiskel, *The Romantic Sublime*. London: Johns Hopkins, 1976: 129.

Bayertz, Kurt. 1985. "Spreading the Spirit of Science: Social Determinants of the Popularization of Science in Nineteenth-Century Germany." In *Expository Science: Forms and Functions of Popularisation*, 209–227. Dordrecht, Neth.: Reidel.

Becker, Lydia Ernestine. 1869. "On the Study of Science by Women." *Contemporary Review* 10: 386–404.

Beeching, H. C., ed. 1925. *The Poetical Works of John Milton.* London: Oxford University Press.

Bell, Susan Groag, and Marilyn Yalom. 1990. Introduction to *Revealing Lives: Autobiography, Biography, and Gender,* ed. Susan Groag Bell and Marilyn Yalom. Albany: State University of New York Press.

Benjamin, Marina, ed. 1991. *Science and Sensibility: Gender and Scientific Enquiry, 1780–1945.* Oxford: Blackwell.

———. 1991. "Elbow Room: Women Writers on Science: 1790–1840." In *Science and Sensibility: Gender and Scientific Enquiry, 1780–1945,* 27–59. Oxford: Blackwell.

Benstock, Shari. 1988. Introduction to *The Private Self: Theory and Practice of Women's Autobiographical Writings.* Chapel Hill: University of North Carolina Press.

Blacker, Carmen. 1996. Preface to *Cambridge Women: Twelve Portraits.* Cambridge, U.K.: Cambridge University Press.

Blair, Hugh Smith. 1763. From *A Critical Dissertation on the Poems of Ossian.* In *The Sublime: A Reader in Eighteenth-Century Aesthetic Theory,* ed. Andrew Ashfield and Peter de Bolla, 207–216. New York: Cambridge University Press, 1996.

Bon-Brenzoni, Caterina. 1853. *I Cieli a Mrs. Mary Somerville.* Milan: Villardi.

Brewster, David. 1829. Letter to John David Forbes, September 11. James David Forbes Papers. St. Andrews University Library.

[Brewster, David.] 1834. Review of *On the Connexion of the Physical Sciences. Edinburgh Review* 119: 154–171.

———. 1837. Review of *History of the Inductive Sciences, from the Earliest to the Present Times. Edinburgh Review* 66: 110–151.

———. 1860. *Memoirs of the Life, Writings, and Discoveries of Sir Isaac Newton.* Vol. 2. Edinburgh: T. Constable and Co.

[Bunbury, E. H.]. 1881. Review of *Life, Letters, and Journals of Sir Charles Lyell. Quarterly Review* 153: 96–131.

Bunders, Joske, and Richard Whitley. 1985. In "Popularisation Within the Sciences: The Purposes and Consequences of Inter-Specialist Communication," 61–77. Dordrecht, Neth.: Reidel.

Burke, Edmund. 1759. "A Philosophical Enquiry into the Origins of Our Ideas of the Sublime and the Beautiful." In *The Sublime: A Reader in Eighteenth-Century Aesthetic Theory,* ed. Andrew Ashfield and Peter de Bolla. New York: Cambridge University Press, 1996.

Byrne, Muriel St. Clare, and Catherine Hope Mansfield. 1921. *Somerville College 1879–1921.* Oxford: Oxford University Press.

Byron, George Gordon. 1905. "Cain, A Mystery," *The Complete Poetical Works of Lord Byron.* In *Cambridge Edition of the Poets,* ed. Paul Elmer More, 621–657. Boston: Houghton Mifflin.

Canning, John, ed. 1983. *100 Great Nineteenth-Century Lives.* London: Methuen.

Cannon, Susan Faye. 1978. *Science in Culture: The Early Victorian Period.* Kent: Science History.

Cantor, Geoffrey. 1996. "The Scientist as Hero: Public Images of Michael Faraday." In *Telling Lives*, ed. Michael Shortland and Richard Yeo, 171–194. Cambridge, U.K.: Cambridge University Press.

Carroll, Berenice. 1990. "The Politics of 'Originality': Women and the Class System of the Intellect." *Journal of Women's History* 2 (Fall): 136–163.

Cawood, John. 1979. "The Magnetic Crusade: Science and Politics in Early Victorian Britain." *Isis* 70: 493–518.

Chambers Journal. 1874. Review of *Personal Recollections*. January 17: 33–36.

Chambers, William. 1878. *Stories of Remarkable Persons*. Edinburgh: Chambers.

Chapone, Mrs. Hester (Mulso). 1772. *Letters on the Improvement of the Mind*. Boston: Wells & Wait.

Clerke, Ellen Mary. 1950. "Somerville, Mary." *Dictionary of National Biography*, 662–663. London: Cumberledge; Oxford: Oxford University Press.

Cloître, Michel, and Terry Shinn. 1985. "Expository Practice: Social, Cognitive and Epistemological Linkage." In *Expository Science: Forms and Functions of Popularisation*, 31–60. Dordrecht, Neth.: Reidel.

Cobbe, Francis Power. 1872. "Blessed Old Age." Obituary of Mary Somerville. *The Echo*. December 3.

1874a. Review of *Personal Recollections*. *Quarterly Review*: 74–103.

1874b. Review of *Personal Recollections*. *Academy*: 1–2.

1878. "Wife Torture in England." *Contemporary Review* 13: 55–87.

1894. *Life of Frances Power Cobbe*. 2 vols. Boston: Houghton Mifflin.

1889. *The Modern Rack: Papers on Vivisection*. London: Sonnenschein.

Cohen, I. Bernard. 1968. "The French Translation of Isaac Newton's *Philosophiae naturalis principia mathematica* (1756, 1759, 1966)." *Archives internationales d'histoire des sciences* 21: 261–290.

Collini, Stefan. 1993. Introduction to *The Two Cultures* by C. P. Snow. Cambridge, U.K.: Cambridge University Press.

Corbett, Mary Jean. 1992. *Representing Femininity: Middle-Class Subjectivity in Victorian and Edwardian Women's Autobiographies*. New York: Oxford University Press.

Creese, Mary R. S. 1991. "British Women of the Nineteenth and Early Twentieth Centuries Who Contributed to Research in the Chemical Sciences." *British Journal for the History of Science* 24: 275–305.

Crosland, Maurice. 1976. "The Development of a Professional Career in Science in France." In *The Emergence of Science in Western Culture*, 139–159. New York: Science History.

Crosland, Maurice, and Crosbie Smith. 1978. "The Transmission of Physics from France to Britain: 1800–1840." *Historical Studies in the Physical Sciences* 9: 1–61.

Dale, Peter Allan. 1989. *In Pursuit of a Scientific Culture: Science, Art, and Society in the Victorian Age*. Madison: University of Wisconsin Press.

Dampier, William Cecil. 1930. *A History of Science and Its Relations with Philosophy and Religion*. Cambridge, U.K.: Cambridge University Press.

Datta, Jean. 1991. "Addressing a Worldwide Readership Through the Filter of Translation." *IEEE Transactions on Professional Communication* 34: 147–150.

David, Deirdre. 1987. *Intellectual Women and Victorian Patriarchy: Harriet Martineau, Elizabeth Barrett Browning, George Eliot*. Ithaca, NY: Cornell University Press.

Davies, Emily. 1988. *The Higher Education of Women*. London: Hambledon.

Davis, Herman S. 1890. "Women Astronomers (1750–1890)." *Popular Astronomy* (June): 211–219.

Dean, Dennis R. 1981. "'Through Science to Despair': Geology and the Victorians." In *Victorian Science and Victorian Values*, ed. James Paradis and Thomas Postlewait. New York: New York Academy of Science.

Dettelbach, Michael. 1996. "Humboldtian Science." In *Cultures of Natural History*, ed. N. Jardine, J. A. Secord, and E. C. Spary, 287–304. New York: Cambridge University Press.

Dryden, John. 1962. "Lines on Milton (1688)." In *Dryden*, ed. Reuben A. Brower. Vol. 104 of *The Laurel Poetry Series*. New York: Dell.

Edgeworth, Maria. 1971. *Letters from England, 1813–1844*, ed. C. Colvin. Oxford: Clarendon.

Findlen, Paula. 1995. "Translating the New Science: Women and the Circulation of Knowledge in Enlightenment Italy." *Configurations* 2: 167–206.

French, Richard D. 1975. *Antivivisection and Medical Science in Victorian Society*. Princeton, NJ: Princeton University Press.

[Galloway, Thomas]. 1832. Review of *Mechanism of the Heavens*. *Edinburgh Review* 109: 1–25.

Gates, Barbara T. 1997. "Revisioning Darwin with Sympathy: Arabella Buckley." In *Natural Eloquence*, ed. Barbara Gates and Ann B. Shteir, 164–178. Madison: University of Wisconsin Press.

Gates, Barbara T., and Ann B. Shteir, eds. 1997. *Natural Eloquence: Women Reinscribe Science*. Madison: University of Wisconsin Press.

Gates, Barbara T., and Ann B. Shteir. 1997. "Introduction: Charting the Tradition." In *Natural Eloquence*, ed. Barbara T. Gates and Ann B. Shteir, 3–24. Madison: University of Wisconsin Press.

"Interview with Diane Ackerman, 18 July 1994." 1997. In *Natural Eloquence*, ed. Barbara T. Gates and Ann B. Shteir, 255–264. Madison: University of Wisconsin Press.

Golinski, Jan. 1992. *Science as Public Culture: Chemistry and Enlightenment in Britain 1760–1820*. New York: Cambridge University Press.

Gramsci, Antonio. 1971. *Selections from the Prison Notebooks of Antonio Gramsci*, ed. and trans. Quintin Hoare and Geoffrey Nowell Smith. New York: International Publications.

Grosser, Morton. [1962] 1979. *The Discovery of Neptune*. New York: Dover.

Hahn, Roger. 1967. *Laplace as a Newtonian Scientist*. Los Angeles: Clark Memorial Library, University of California.

Hale, Mary W. 1840. "Comparative Intellectual Character of the Sexes." *Godey's Lady's Book* 20: 273–275.

Hall, Spencer. 1997. "Feminism, Ecology, Romanticism." Review of *Becoming Wordsworthian: A Performance Aesthetics*, by Elizabeth A. Fay, and *Ecological Literary Criticism: Romantic Imagining and the Biology of Mind*, by Karl Kroeber. *College English* 59: 828–829.

Hamer, Sarah Sharp. 1887. *Mrs. Somerville and Mary Carpenter*. London: Cassell.

Harvey-Gibson, R. J. 1929. *Two Thousand Years of Science*. New York: Macmillan.

Hays, J. N. 1964. "Science and Brougham's Society." *Annals of Science* 20: 227–241.

Helsinger, Elizabeth K., Robin Lauterbach Sheets, and William Veeder. 1983. *The Woman Question: Defining Voices, 1837–1883*. Vol. 1 of *The Woman Question: Society and Literature in Britain and America, 1837–1883*. New York: Garland.

Helsinger, Elizabeth K., Robin Lauterbach Sheets, and William Veeder. 1983. *The Woman Question: Social Issues, 1837–1883*. Vol. 2 of *The Woman Question: Society and Literature in Britain and America, 1837–1883*. New York: Garland.

[Herschel, J. F. W.]. 1832. Review of *Mechanism of the Heavens. Quarterly Review* 47: 537–559.

——— 1840. *Preliminary Discourse on the Study of Natural Philosophy (1830)*. New York: Harper.

——— 1841. Review of *History of the Inductive Sciences from the Earliest to the Present Times. Quarterly Review* 68: 177–238.

Heyck, T. W. 1982. *The Transformation of Intellectual Life in Victorian England*. Chicago: Lyceum.

Hill, John Spencer. 1977. Introduction to *The Romantic Imagination: A Casebook*. London: Macmillan.

[Holland, Henry]. 1848. Review of *Physical Geography. Quarterly Review* 83: 305–340.

Hollis, Patricia. 1979. *Women in Public, 1850–1900: Documents of the Victorian Women's Movement*. London: Allen.

Humboldt, Alexander von. 1850. *Cosmos: A Sketch of a Physical Description of the Universe*, trans. E. C. Otté. 2 vols. New York: Harper and Brothers.

Jardine, N., J. A. Secord, and E. C. Spary, eds. 1996. *Cultures of Natural History*. New York: Cambridge University Press.

Jardine, N., and E. C. Spary. 1996. "The Natures of Cultural History." In *Cultures of Natural History*, ed. N. Jardine, J. A. Secord, and E. C. Spary, 3–13. Cambridge: Cambridge University Press.

Johnson, W. 1994. "Some Women in the History of Mathematics, Physics, Astronomy and Engineering." *Journal of Materials Processing Technology* 40: 33–71.

Jones, William Powell. 1959. "Science in Biblical Paraphrases in Eighteenth-Century England." *Publications of the Modern Language Association of America* 74: 41–51.

——— 1961. "The Idea of the Limitations of Science from Prior to Blake." *Studies in English Literature 1500–1900* 1: 97–114.

——— 1966. *The Rhetoric of Science: A Study of Scientific Ideas and Imagery in Eighteenth-Century English Poetry*. Berkeley: University of California Press.

Jordanova, Ludmilla. 1993. "Gender and the Historiography of Science." *British Journal for the History of Science* 26: 469–483.

Keller, Evelyn Fox. 1995. "Gender and Science: Origin, History, and Politics."
     Osiris 10: 27–38.
Kellner, L. 1963. *Alexander von Humboldt*. London: Oxford University Press.
Kelso Mail. 1874. Review of *Personal Recollections*. March 24.
Kendall, Phebe Mitchell, ed. 1896. *Maria Mitchell: Life, Letters, and Journals*.
     Boston: Lee.
Kohlstedt, Sally Gregory. 1987. "Maria Mitchell and the Advancement of
     Women in Science." In *Uneasy Careers and Intimate Lives: Woman in
     Science 1789–1979*, ed. Pnina Abir-Am and Dorinda Outram, 129–146.
     New Brunswick, NJ: Rutgers University Press.
———. 1978. "In from the Periphery: American Women in Science, 1830–1880."
     Signs 4: 81–96.
———. 1995. "Women in the History of Science: An Ambiguous Place." *Osiris* 10:
     39–58.
Kohlstedt, Sally Gregory, and Helen E. Longino, eds. 1997. "The Women,
     Gender, and Science Question: What Do Research on Women in
     Science and Research on Gender and Science Have to Do with Each
     Other?" *Osiris* 12: 3–15.
Kuhn, Thomas. 1962. "Energy Conservation as an Example of Simultaneous
     Discovery." *Critical Problems in the History of Science*, ed. Marshall
     Clagett, 321–356. Madison, WI: University of Wisconsin Press.
Lansbury, Coral. 1985. *The Old Brown Dog: Women, Workers, and Vivisection*.
     Madison: University of Wisconsin Press.
Laplace, Pierre Simon Marquis de. 1829. *Traité de Mécanique Céleste*. Second
     ed. 2 vols. Paris: Bachelier.
Laudan, Rachel. 1993. "Histories of the Sciences and Their Uses: A Review to
     1913." *History of Science* 31: 1–34.
Lear, Linda. 1997. *Rachel Carson: Witness for Nature*. London: Allen Lane.
Lightman, Bernard. 1987. *The Origins of Agnosticism: Victorian Unbelief and the
     Limits of Knowledge*. Baltimore: Johns Hopkins.
———. 1997. "Constructing Victorian Heavens: Agnes Clerke and the 'New
     Astronomy.'" In *Natural Eloquence*, ed. Barbara T. Gates and Ann B.
     Shteir, 61–78. Madison: University of Wisconsin Press.
Lindee, M. Susan. 1996. "The American Career of Jane Marcet's *Conversations
     on Chemistry*, 1806–1853." In *Technical Knowledge in American Culture:
     Science, Technology, and Medicine Since the Early 1800s*, ed. Hamilton
     Cravens, Alan Marcus, and David M. Katzman, 40–52. Tuscaloosa:
     University of Alabama Press.
"Literature of the Day." 1874. Review of *Personal Recollections*. *Lippincott's
     Magazine* 13: 518–520.
Lloyd, Genevieve. 1984. *The Man of Reason: "Male" and "Female" in Western
     Philosophy*. London: Methuen.
Mason, Joan. 1992. "The Admission of the First Women to the Royal Society
     of London." *Notes and Records of the Royal Society at London* 46: 279–
     300.
———. 1995. "The Women Fellows' Jubilee." *Notes and Records of the Royal Society
     at London* 49: 125–140.

1996. "Grace Chisholm Young and the Division of Laurels." *Notes and Records of the Royal Society at London* 50: 89–100.

Maxwell, J. C. 1890. *Scientific Papers of James Clerke Maxwell*. 2 vols. Cambridge: Cambridge University Press.

Mellor, Anne K. 1993. *Romanticism and Gender*. New York: Routledge.

Merchant, Carolyn. 1980. *The Death of Nature: Women, Ecology, and the Scientific Revolution*. San Francisco: Harper & Row.

Merz, John Theodore. 1965. *A History of European Thought in the Nineteenth Century*. Vol. 1. New York: Dover.

Morning Post. 1872. Obituary of Mary Somerville. December 2.

Morrell, Jack, and Arnold Thackray. 1981. *Gentlemen of Science: Early Years of the BAAS*. Oxford: Clarendon.

Moseley, Maboth. 1964. *Irascible Genius: A Life of Charles Babbage, Inventor*. London: Hutchinson.

Mozans, H. J. [John Augustine Zahn]. 1913. *Woman in Science, With an Introductory Chapter on Woman's Long Struggle for Things of the Mind*. New York: Appleton.

Myers, Greg. 1985. "Nineteenth-Century Popularizations of Thermodynamics and the Rhetoric of Social Prophecy." *Victorian Studies* 29: 35–66.

1990. *Writing Biology: Texts in the Social Construction of Scientific Knowledge*. Madison: University of Wisconsin Press.

1997. "Fictionality, Demonstration, and a Forum for Popular Science: Jane Marcet's *Conversations on Chemistry*." In *Natural Eloquence*, ed. Barbara T. Gates and Ann B. Shteir, 43–60. Madison: University of Wisconsin Press.

Nature. 1874. "Mary Somerville." Review of *Personal Recollections*. April 2: 417–418.

Neeley, Kathryn A. 1992. "Woman as Mediatrix: Women as Writers on Science and Technology in the Eighteenth and Nineteenth Centuries." *IEEE Transactions on Professional Communication* 35: 208–217.

New York Herald. 1874. "Ideal Old Age." Rev. of *Personal Recollections*. January 4.

Nicolson, Marjorie Hope. 1946. *Newton Demands the Muse: Newton's Opticks and the Eighteenth-Century Poets*. Princeton, NJ: Princeton University Press.

1959. *Mountain Gloom and Mountain Glory: The Development of the Aesthetics of the Infinite*. Ithaca, NY: Cornell University Press.

Niven, W. D., ed. 1890. *The Scientific Papers of James Clerk Maxwell*. 2 vols. Cambridge, U.K.: Cambridge University Press.

Ogilvie, Marilyn Bailey. 1986. *Women in Science: Antiquity through the Nineteenth Century*. Cambridge, MA: MIT Press.

Oldroyd, David. 1986. *The Arch of Knowledge: An Introductory Study of the History of the Philosophy and Methodology of Science*. New York: Methuen.

Osen, Lynn N. 1974. *Women in Mathematics*. Cambridge, MA: MIT Press.

Outram, Dorinda. 1987. "Before Objectivity: Wives, Patronage, and Cultural Reproduction in Early Nineteenth-Century French Science." In *Uneasy Careers and Intimate Lives: Women in Science 1789–1979*, ed. Pnina Abir-

Am and Dorinda Outram, 19–30. New Brunswick, NJ: Rutgers University Press.

Paradis, James, and Thomas Postlewait, eds. 1981. *Victorian Science and Victorian Values: Literary Perspectives*. New York: New York Academy of Sciences.

Parker, Rozsika, and Griselda Pollack. 1981. *Old Mistresses: Women, Art, and Ideology*. London: Routledge.

Patterson, Elizabeth Chambers. 1969. "Mary Somerville." *British Journal for the History of Science* 4: 311–339.

    1974. "The Case of Mary Somerville: An Aspect of Nineteenth-Century Science." *Proceedings of the American Philosophical Society* 118: 269–275.

    1975. "Somerville, Mary Fairfax Greig." In *Dictionary of Scientific Biography*, ed. Charles Coulston Gillispie. New York: Scribner.

    1979. *Mary Somerville, 1780–1872*. Oxford: Bocardo.

    1983. *Mary Somerville and the Cultivation of Science, 1815–1840*. The Hague: Martinus Nijoff.

Perl, Teri. 1978. *Math Equals: Biographies of Women Mathematicians + Related Activities*. Menlo Park, CA: Addison-Wesley.

Peterson, M. Jeanne. 1989. *Family, Love, and Work in the Lives of Victorian Gentlewomen*. Bloomington: Indiana University Press.

Phillips, Patricia. 1990. *The Scientific Lady: A Social History of Women's Scientific Interests 1520–1918*. New York: St. Martin's Press.

Playfair, John. 1819. *Outlines of Natural Philosophy*. 3rd ed. Edinburgh: Constable.

Poovey, Mary. [1988] 1991. *Uneven Developments: The Ideological Work of Gender in Mid-Victorian England*. London: Virago.

Pope, Alexander. 1969. "Epitaph Intended for Sir Isaac Newton (1730)." In *Eighteenth-Century English Literature*, ed. Geoffrey Tillotson, Paul Fussell, Jr., and Marshall Waingrow. New York: Harcourt, Brace, and World.

Preyer, Robert O. 1981. "The Romantic Tide Reaches Trinity: Notes on the Transmission and Diffusion of New Approaches to Traditional Studies at Cambridge, 1820–1840." In *Victorian Science and Victorian Values*, ed. James Paradis and Thomas Postlewait, 39–110. New York: New York Academy of Sciences.

Proctor, Richard. 1873. *Light Science for Leisure Hours, Second Series: Familiar Essays on Scientific Subjects, Natural Phenomena, &c. With a Sketch of the Life of Mary Somerville*. London: Longmans.

Pycior, Helena M. 1987. "Marie Curie's 'Anti-Natural Path': Time Only for Science and Family." In *Uneasy Careers and Intimate Lives: Women in Science 1789–1979*, ed. Pnina Abir-Am and Dorinda Outram, 191–215. New Brunswick, NJ: Rutgers University Press.

Pycior, Helena M., Nancy G. Slack, and Pnina Abir-Am, eds. 1996. *Creative Couples in the Sciences*. New Brunswick, NJ: Rutgers University Press.

Raglon, Rebecca. 1997. "Rachel Carson and Her Legacy." In *Natural Eloquence*, ed. Barbara T. Gates and Ann B. Shteir, 196–214. Madison: University of Wisconsin Press.

Richeson, A. W. 1941. "Mary Somerville." *Scripta Mathematica* 8: 5–13.

Ritvo, Harriet. 1987. *The Animal Estate: The English and Other Creatures in the Victorian Age*. Cambridge, MA: Harvard University Press.

Ronan, Colin A. 1983. *The Cambridge Illustrated History of the World's Science*. New York: Cambridge University Press.

[Roscoe, H. E.] 1869. Review of *Molecular and Microscopic Science*. Mary Somerville. *Edinburgh Review* 130: 70–83.

Rose, Hilary. 1994. *Love, Power, and Knowledge: Towards a Feminist Transformation of the Sciences*. Cambridge, U.K.: Polity Press.

Rossiter, Margaret. 1987. Foreword to *Uneasey Careers and Intimate Lives: Women in Science 1789–1979*, ed. Pnina Abir-am and Dorinda Outram, xi–xii. New Brunswick, NJ: Rutgers University Press.

———. 1982. *Women Scientists in America: Struggles and Strategies to 1940*. Baltimore: Johns Hopkins University Press.

Russett, Cynthia. 1989. *Sexual Science: The Victorian Construction of Womanhood*. Cambridge, MA: Harvard University Press.

The Saturday Review. 1874. "Mrs. Somerville." Review of *Personal Recollections*. January 10: 53–55.

Sayers, Dorothy L. 1968. *Gaudy Night*. New York: Avon.

Schiebinger, Londa. 1989. *The Mind Has No Sex? Women in the Origins of Modern Science*. Cambridge, MA: Harvard University Press.

———. 1989. *Nature's Body: Women in the Origins of Modern Science*. Boston: Beacon Press.

Schweber, S. S. 1981. "Scientists as Intellectuals: The Early Victorians." In *Victorian Science and Victorian Values*, ed. James Paradis and Thomas Postlewait, 1–37. New York: New York Academy of Sciences.

Secord, James A. "Crisis of Nature." In *The Cultures of Natural History*, ed. N. Jardine, J. A. Secord, and E. D. Spary, 447–459. New York: Cambridge University Press.

Sedgwick, W. T., and H. W. Tyler. 1919. *A Short History of Science*. New York: Macmillan.

Shea, William R. 1994. "Galileo in the Nineties." *Perspectives on Science* 2: 476–487.

Shepherd, Linda Jean. 1993. *Lifting the Veil: The Feminine Face of Science*. Boston: Shambhala.

Shils, Edward, and Carmen Blacker, eds. 1996. *Cambridge Women: Twelve Portraits*. Cambridge, U.K.: Cambridge University Press.

Shinn, Terry, and Richard Whitley, eds. 1985. *Expository Science: Forms and Functions of Popularisation*. Dordrecht, Neth.: Reidel.

Shortland, Michael, and Richard Yeo, eds. 1996. *Telling Lives in Science: Essays on Scientific Biography*. Cambridge, U.K.: Cambridge University Press.

Shortland, Michael and Richard Yeo. 1996. Introduction to *Telling Lives in Science: Essays on Scientific Biography*, ed. Michael Shortland and Richard Yeo, 1–44. Cambridge, U.K.: Cambridge University Press.

[Simcox, Edith.] 1874. Review of *Personal Recollections*. *Fortnightly Review* 21: 109–120.

Singer, Charles J. 1959. *A Short History of Scientific Ideas to 1900*. Oxford: Clarendon.

Smith, Jonathan. 1994. *Fact and Feeling: Baconian Science and the Nineteenth-Century Literary Imagination*. Madison: University of Wisconsin Press.

Söderqvist, Thomas. "Existential Projects and Existential Choice in Science: Science Biography as Edifying Genre." In *Telling Lives*, ed. Michael Shortland and Richard Yeo, 45–102. Cambridge, U.K.: Cambridge University Press.

Somerville Collection. Bodleian Library. Somerville College, Oxford.

Somerville, Martha. 1874. *Personal Recollections from Early Life to Old Age of Mary Somerville, with Selections from her Correspondence. By her daughter, Martha Somerville (1873)*. Boston: Roberts.

Somerville, Mary. 1826. "On the Magnetizing Power of the More Refrangible Solar Rays." *Philosophical Transactions of the Royal Society of London* 116: 132–139.

———. 1831. *Mechanism of the Heavens*. London: Murray.

———. 1832. *A Preliminary Dissertation on the Mechanism of the Heavens*. Philadelphia: Carey & Lea.

———. 1835. "The Comet." *Quarterly Review* 55: 195–223. (Full title: "Art. VII. – 1. Ueber den Halleyschen Cometen. Von Littrow. Wien, 1835. 2. Ueber den Halleyschen Cometen. Von Professor von Encke. Berline Jahrbuch, 1835 &c. &c. &c.")

———. 1836. "Experiences sur la transmission des rayons chimiques du spectre solaire, à travers différents milieux. Extrait d'une lettre de Mme Sommerville [sic] à M. Arago." *Comptes Rendus* 3: 473–476.

———. 1845. " 'On the Action of the Rays of the Spectrum on the Vegetable Juices': being an Extract from a Letter by Mrs. M. Somerville to Sir John F. W. Herschel, Bart., dated Rome, September 20, 1845. Communicated by Sir John F. W. Herschel, Bart., F. R. S." *Abstracts of the Papers Communicated to the Royal Society of London from 1843 to 1850* 5: 569–570.

———. 1845. "On the Action of the Rays of the Spectrum on the Vegetable Juices. Extract of a Letter from Mrs. M. Somerville to Sir J. F. W. Herschel, Bart., dated Rome, September 20, 1845. Communicated by Sir J. Herschel." *Philosophical Transactions of the Royal Society of London* 136: 111–120.

———. 1846. *On the Connection of the Physical Sciences (1834)*. 7th London ed. New York: Harper and Brothers.

———. 1850. *Physical Geography (1848)*. Phildelphia: Lea & Blanchard.

———. 1869. *On Molecular and Microscopic Science*. London: Murray.

———. n.d. First manuscript draft of *Personal Recollections*. Somerville Collection. Folder MSAU-2 in Dep.c.355.

———. n.d. Second manuscript draft of *Personal Recollections*. Somerville Collection. Folder MSAU-3 in Dep.c.355.

Sonnert, Gerhard. 1995. *Who Succeeds in Science? The Gender Dimension*. New Brunswick, NJ: Rutgers University Press.

Sotheby, William. 1834. *Lines Suggested by the Third Meeting of the British Association for the Advancement of Science*. London: Nicol.

Soudek, Ingrid H., and Kathryn A. Neeley. 1994. "The Match and Other Agents of Liberation: The Role of Technology in the Social Thought of Louise Otto." In *Research in Philosophy and Technology: Technology and Everyday Life*, ed. George Allan. Vol. 14: 119–137. Greenwich, Connecticut: JAI Press.

Spectator. 1872. Obituary of Mary Somerville. December 7.

Spectator. 1874. Review of *Personal Recollections*. January 31: 143.

Tabor, Margaret. 1933. *Pioneer Women: Caroline Herschel, Sarah Siddons, Maria Edgeworth, Mary Somerville*. 4th series. London: Sheldon Press.

Tennyson, Alfred. 1989. "The Princess." *Tennyson: A Selected Edition*, ed. Christopher Ricks. Berkeley: University of California Press.

The Times. 1922. "University News: Cambridge, Nov. 3." November 4: 7e.

Todhunter, Isaac. 1876. *William Whewell, D.D.: An Account of His Writings with Selections from his Literary and Scientific Correspondence*. London: Macmillan.

Tomaselli, Sylvana. 1988. "Collecting Women: The Female in Scientific Biography." *Science as Culture* 4: 95–106.

Tomkins, Lydia [Henry Brougham]. 1835. *Thoughts on the Ladies of the Aristocracy*. London: Hodgsons.

Topham, Jonathan. 1992. "Science and Popular Education in the 1830s: The Role of the *Bridgewater Treatises*." *British Journal of the History of Science* 25: 397–430.

Toth, Bruce, and Emily Toth. 1978. "Mary Who?" *Johns Hopkins Magazine* 29: 25–29.

Weiskel, Thomas. [1976] 1986. *The Romantic Sublime*. London: Johns Hopkins.

Whewell, William. 1834. Review of *On the Connexion of the Physical Sciences*. *Quarterly Review* 51: 54–68.

——— 1837a. *On the Principles of English University Education*. London: John W. Parker.

——— 1837b. *History of the Inductive Sciences*. London: John Parker.

Wickens, G. Glen. 1980. "The Two Sides of Early Victorian Science and the Unity of 'The Princess.'" *Victorian Studies* 23: 369–388.

Williams, Henry Smith. 1901. *The Story of Nineteenth-Century Science*. New York: Harper and Brothers.

Yearley, Steven. 1985. "Representing Geology: Textual Structures in the Pedagogical Presentation of Science." In *Expository Science: Forms and Functions of Popularisation*, ed. Terry Shinn and Richard Whitley, 79–101. Dordrecht, Neth.: Reidel.

Yeo, Richard. 1987. "William Whewell on the History of Science." *Metascience* 5: 25–40.

——— 1993. *Defining Science: William Whewell, Natural Knowledge, and Public Debate in Early Victorian Britain*. New York: Cambridge University Press.

Zaniello, Tom. 1988. *Hopkins in the Age of Darwin*. Iowa City: University of Iowa Press.

Index